Common Sense Physics & Cosmology

Never before, a comprehensive logical interpretation of the universe including all phenomena is discovered, which is consistent and coherent supported by well-established findings cumulated over several centuries! It has to be the only correct physics and cosmology and it is no longer possible to have an alternative interpretation of the universe also supported by all findings.

Since the interpretation is based on the classical Galilean physics, it is proven to be the only correct physics and cosmology!

Josef Tsau, Ph.D.

ISBN 0-7414-5773-3

Published by:

INFINITY
PUBLISHING.COM

1094 New DeHaven Street, Suite 100
West Conshohocken, PA 19428-2713
Info@buybooksontheweb.com
www.buybooksontheweb.com
Toll-free (877) BUY BOOK
Local Phone (610) 941-9999
Fax (610) 941-9959

Printed in the United States of America

Published September 2010

To the only correct nature science and my parents

This book is dedicated to the only correct nature science to benefit mankind. It is also for the memory of my parents, who have brought me up to be a dedicated scientist.

Preface

Twelve years has past since the author devoted himself to solve physics problems. It began with his idea that only neutrinos could produce such universal phenomena as gravity, light etc. He soon found that unlike other fields of the natural science physics was still full of controversial ideas and theories. Further to his surprise that his concept was unacceptable by the so-called mainstream of thought, the worldwide accepted theories in the past century teaching mathematical origins instead of physical origins but only for the interpretations of the universe and universal phenomena.

The author continued to find his idea promising but he had to build his universe alone based on classical Galilean physics since his view has been repeatedly rejected by physics journals and NSF and ignored by major university professors and officers. The steady progress of his searching for an overall understanding of the universe is shown from his published books such as the "Gravity, Light, and the Universe", 2001; the "Discovery of Aether and its Science" in 2005; and the "Real Physics and Cosmology" in 2007. Now, the author is happy to finally offer a logical overall understanding of the universe and all universal phenomena, which is consistent, coherent, and is supported by well established R&D findings cumulated over several centuries.

The nature, not the church, the scientific community, or the government, is the final authority to tell us what the only correct natural science is! The nature has been teaching us for centuries that only the correct physics can offer an overall understanding of the nature supported by all findings. Now, it should be very clear particularly to scientists that Galilean physics has to be the only correct physics and cosmology, that all theory incompatible with Galilean physics such as the mainstream of thought have been proven wrong, and that it is not possible to come up with an alternative interpretation of the universe also supported by all findings. Now, it is simply invalid for any authority to reject that Galilean physics is the proven correct physics using any criterion such as those established by the mainstream of thought and by any authority such as the physics society or the government.

This book therefore should be the unique text book for all, scientists, engineers, teachers, students, and the public, to re-learn physics and cosmology offering an exiting overall advanced scientific understanding while pointing out past mistakes and misunderstandings. Without complicated mathematics, physics and cosmology are now easy and fun to learn again.

Galilean physics is the part of the classical physics that has repeatedly proven correct and it teaches that all phenomena have physical origins, which are matters such as material particles having masses and relative motions. Long-time failure in understanding and in finding the physical origins for such important universal phenomena as charge energy, gravity, light, etc. has led to the development of theories using mathematical equations to interpret and to represent them. The mainstream of thought including both modern physics and modern cosmology based on Einstein's theories of relativity is such a mathematical theory.

The breakthrough discovery given in this book shows that both the universal phenomena and the universe can be logically interpreted by Galilean physics alone without those mathematical theories such as the force-field theories of Faraday, the electromagnetic wave theory of Maxwell, the theories of relativity of Einstein, the big bang theory, and their postulations. They all therefore fall to Ocam's Razor! The only correct physics should be the physics of everything capable of interpreting the universe and all phenomena consistently and coherently supported by all experimental findings cumulated over several centuries. In the past century, the mainstream of thought has enjoyed the huge advantage being exclusively supported by concentrated R&D effort worldwide, when

Einstein and many others have tried to develop a theory of everything seeking mathematical origins to interpret the universe but ended with frustrations. Now, there is no doubt that the race to find the only correct physics is over.

This book continues to lead the most important scientific revolution: to replace the mainstream of thought by the newly discovered expanded Galilean physics and cosmology given in it. It has long been the governments of all nations leading and financing the R&D of natural science particularly for physics and cosmology. In the past century they all have been supporting the mainstream of thought and have established an authoritative system using its knowledge and formal-science philosophy, which have not been prove correct scientifically, as the scientific standards to block opposing scientific views and R&D work. Many scientists have long openly suggested that the mainstream of thought is wrong and is religion-like. In fact, the world has been in an era of mathematically created pseudo-physics and pseudo-cosmology for a century already amazingly similar to the ancient geocentric era, which had lasted over a thousand years.

The governments of all nations would not intentionally support wrong physics to hurt their people and the future. For example, it is against law for U. S. Government to support religion and religion-like pseudo-science such as the teaching of Creation in schools. It is time to hold our government responsible to teach and to support the only correct physics and cosmology and to sponsor valid scientific evaluation of the newly discovered expanded Galilean physics against all mathematic theories such as the mainstream of thought.

Now the author has proven that Galilean physics is the only correct physics and cosmology and challenges the world with his proof.

Table of Contents

Chapter 1 –The only correct physics

Nature science and its scientific method

Physics is part of the natural science, which finds the laws of the nature. The laws of nature can neither be changed nor be man-made. Scientific method uses experimental findings and observations to discover them. Scientific method also includes using logic thinking to interpret findings and observations based on well established scientific knowledge of natural laws, to figure out their technological applications, and to discover new natural laws. All animals and even plants have the ability to adopt some laws of the nature to survive. Only human has the most powerful brain of logical thinking to cumulatively gather the knowledge of natural laws making technological applications. Since there can be only one correct scientific answer, natural science must be consistent and coherent. The scientific knowledge therefore must be cumulative to maintain consistency and coherency. Scientific method therefore further involves obeying proven-correct scientific laws and knowledge and using them as the guide to advance natural science with consistency and coherency.

Due to the cumulative nature, it is most important that the foundation of natural science is correct. Otherwise, the entire natural science so based would be wrong. For example, the well known geocentric cosmology before 17th Century was based on religious belief (unscientific postulation) that Earth was the center of the universe. It had lasted over a thousand years before it was proven wrong because it was religion-like. To advance science consistently, even proven-correct scientific laws and knowledge must be repeatedly challenged by new findings but must also be obeyed unless they are proven wrong. Well-established scientific laws and knowledge therefore are those repeatedly tested by new findings and applications for a long period of time to assure their correctness. The postulations of a scientific theory therefore must be consistent with well-established laws and knowledge to be acceptable. Scientists therefore should be guided by the well established knowledge and laws of natural science.

The only correct natural science

The most important question is, therefore, what the only correct natural science is? The best answer is that it must be supported by all experimental findings and observations and therefore it must be the natural science of everything. It has been over four centuries since the scientific revolution of Copernicus and Galileo to clearly define natural science and to separate it from religion and an amazingly huge amount of experimental findings and observations by worldwide R&D have been generated and cumulated ever since. Since only the correct natural science can be supported by all the cumulated findings, there has long been little doubt on what is the real natural science in almost all fields of the natural science except physics and cosmology. Instead of protecting their own theories, it should be time for the communities of both physics and astronomy to face the centuries-long cumulated findings together to determine what the only correct physics and cosmology are.

Many experimental findings may be interpreted differently and all valid scientific interpretations must be considered. For example, the well established scientific knowledge is that nuclear reactions release huge amount of energy and lose a fraction of mass. The worldwide accepted interpretation is that the lost mass turns into huge amounts of (pure) energy (obeying Einstein Equation). This interpretation however contradicts the well-established mass-balance law established from the knowledge of both physical and chemical reactions. The findings however can also be interpreted as: the mass lost in all nuclear reactions is due to the release of mass-containing particles such as neutrinos in these reactions, which are too small to be detectable and the huge amount of energy is produced by the released particles. Having not considered or unscientifically ruling out this possible interpretation, the above accepted interpretation only has 50% chance of being correct. Worse yet, to be correct, it should be consistent with Galilean physics, the well established scientific knowledge, but it is not. Galilean physics teaches that all energies should be produced by matters having masses and relative motions. The "pure energy" is scientifically indefinable by Galilean physics. As a result, the above accepted interpretation of the nuclear reactions is likely wrong and the proposed new scientific interpretation consistent with Galilean physics should be considered!

Physics is unique in natural science still having several contradictory views or theories coexisting. Besides Galilean physics, there is the mainstream of thought, which includes both the modern physics and the modern cosmology developed in the past century based on Einstein's theories of relativity inconsistent with Galilean physics. There also other theories supported by many physicists, such as the plenum aether theory, which are inconsistent with both Galilean physics and the mainstream of thought.

There however can be only one correct physics capable of interpreting everything and all phenomena consistently and coherently. Einstein and many others have been trying to develop a theory of everything without success. Therefore, the race is still on to find the only correct physics.

Galilean physics

Core of physics

Besides his important contribution to astronomy, Galileo has led to the development of physics of mechanics (PM), the so-called Galilean physics. Physics therefore is considered to begin from Galileo in 17th century. PM studies fundamental physical properties of matters, such as their motions, momentum, energies, and forces. Newton (1642-1727), the famous successor of Galileo, is known for his three laws of motion and his universal theory of gravitation. In 19th Century PM was expanded to cover physics of heat, thermodynamics and kinetics. Physics of heat is the study of energy, energy transfer and heat phenomena. The heat phenomenon has been explained by the motion energy of atoms and molecules and temperature is a measurement of the kinetic energy of atomic and molecular motions. PM has long been the core of physics often referred to as the determinant physics meaning that it is experimentally verifiable and quantifiable. It is also the part of physics that makes common sense emphasizing logical understanding. However, its progress has been stopped from the long-time failure in finding the matters it has predicted to produce such universal phenomena as light and gravity.

PM remains to be the only well-established part of the physics that has been repeated proven correct over the past four centuries essentially without unproved postulation. Even today PM is still the core of physics. We should be very sure that PM is the correct physics. Physics is the foundation of all branches of natural sciences. It is therefore particularly important that physics is correct. A century after the acceptance of the mainstream of thought, all other fields of natural science are still based on PM essentially untouched by the mainstream of thought.

Phenomena having physical origins

PM teaches that all matters have masses, occupy certain amounts of space, and have relative motions. Having these basic physical properties, they produce phenomena such as forces and energies from their collision interactions. Their collisions produce pushing forces to act on each other and to exchange energies resulting in changing their relative motions and energy levels. PM therefore teaches that matters produce all phenomena from their existence and their collision interactions. All matters have some common properties measurable and mathematically definable such as their energies (e.g., $0.5mv^2$), forces (e.g. ma), momentum (e.g., mv), etc. These mathematical definitions also teach that phenomena must be produced by matters having masses meaning that phenomena must have physical origins or produced by matters such as material particles. These mathematical definitions however represent the common properties of all matters not specific to any one of them. Mathematical equations or definitions therefore are incapable of represent any specific matter, such as its composition and other unique properties and, thus, can neither be nor represent the physical origin of any phenomenon. PM therefore teaches that mathematics is only a good tool for natural science.

Since PM has been repeatedly proven correct over the past several centuries, its teachings should be correct and obeyed. It teaches that all phenomena occur for specific reasons and they, such as fire and sound, are produced by matters, which have their own specific physical and chemical properties discoverable only by experiments not by mathematical manipulations. Detecting and measuring properties common to all matters alone cannot find the specific physical origins of a phenomenon. A phenomenon cannot be expressed by a mathematical function and cannot be discovered by mathematical manipulations.

Air particles produce such global phenomena as wind, atmospheric pressure, fire, and sound. PM teaches that universal phenomena, such as universal forces and light, should also be produced by matter(s).

Excluding Newton's theory of gravitation

Newton's three laws of motion are important part of PM but his theory of universal gravitation is not, since his postulations that all matters have acting-from-distance attraction force are incompatible with the teachings of PM, which teaches that there is only acting-on-contact pushing force. Newton's universal theory of gravitation therefore must be excluded from PM.

Standing tall

Experimental findings are the only way to determine what correct physics is. Although some findings have more than one possible interpretation, very large amount of experimental findings always point to the only correct physics. In the presence of very large amount of findings, the only correct physics will stand tall supported by all of them. Over the past four centuries, tremendous amount of experimental findings have been generated, which has repeatedly proven PM correct scientifically, mathematically, and in all technological applications.

The only correct physics should be the physics of everything and the final test is for PM to come up with an interpretation of the universe and all universal phenomena consistently and coherently. Unfortunately, upon supporting mathematical theories including the mainstream of thought, scientists have in fact long given up PM having not given PM a chance to prove that it is the only correct physics. Finally, this book reports the breakthrough discovery of the physical origins of all universal phenomena and the universe, thus, offering an overall interpretation of the universe supported by the well established findings cumulated in the past several centuries. The expanded PM given in this book is therefore proven to be the only correct physics and it should also be the physics of everything.

Chapter 2 – Gravity, strong and weak forces

Discovery of aether

Neutrino particles

To scientists, neutrinos are still the most mysterious matter. To begin with, their existence was postulated, not discovered. In order for the beta-decay process to obey the law of conservation of energy, Wolfgang Pauli in 1930 postulated that this process also released an elusive particle. It was named neutrino and was only indirectly detected 23 years later by Cowen and Reines. Now, scientists know that all stars produce neutrinos from their nuclear reactions and there are three kinds of them. They estimated that there were hundreds of millions of neutrinos per second passing through our body coming from Sun alone. Since there are billions of stars in the universe around us, there is certainly a very dense energetic atmosphere of neutrino particles in the universe. Believing that neutrinos can freely penetrating through all matters, scientists put very large tank filled with heavy water solutions very deep underground to detect and to study neutrinos, where no other cosmic particles can reach. They think that by "rare chance" high-energy neutrinos can hit molecules and break their chemical bonds of the heavy water or dissolved chemicals to produce very faint light. They have installed very sensitive light detectors in the tank to detect the light-producing neutrinos. By finding the surge of energetic neutrinos simultaneously with the detecting a new supernova, scientists have concluded that neutrinos travel at close to light speed. Several recent findings suggested that neutrinos do have very small mass. More detailed description of the findings and the theory of neutrino of the modern physics are given in the Discussion chapter. Neutrinos are such elusive particles that they are still undetectable directly. The methods developed to detect them and the results obtained so far have many problems such as sensitivity, specificity, and interpretation problems.

The postulated tiny particles

Many phenomena, such as sound, fire, atmospheric pressure, wind, the presence of animal and plants, etc. found anywhere on Earth, are mainly or partially due to the presence of an atmosphere of air. There are also many phenomena such as universal forces and light occurring anywhere in the universe. There is likely an atmosphere of some kind of material particles (aether) taking up all the space in the universe to produce them.

Earth does not continue to produce air and the atmosphere of air is kept on the surface of Earth by gravitational force. Without gravity air will quickly escape from Earth. The nature appears to have unlimited space and we may define our universe to be a region where many stars exist. If there is an atmosphere of material particles or aether in the universe, they likely will continuously escape out of the universe. For the universe to constantly have an atmosphere of aether having essentially constant concentration and energy, the universe must continuously produce them at a constant rate to compensate their loss from escaping out of the universe. Besides, there must be many locations about evenly distributed in the universe to produce them in order to have an essentially even distribution of its concentration and energy in the universe.

It has been known for quite sometime that all stars produce neutrinos at constant rates from their nuclear reactions. Although the presence of an atmosphere of neutrinos meets all the above scientific conditions for being the aether, scientists have not found evidence that neutrinos interact with matters (no heat produced) and therefore believe that they are very inactive and that they do not interact with ordinary matters to produce universal phenomena. They even believed that neutrinos have no mass until recently.

Based on the teachings of PM scientists have long predicted the presence of something to produce all universal phenomena. For example, Fatio, Le Sage, and Newton have proposed the presence of tiny particles interacting with matters to produce gravity and light during 16-18th centuries. The properties, such as capable of penetrating through ordinary matters, of the postulated tiny particles by Le Sage are surprisingly similar to those of the neutrinos!

NEU

To be able to produce gravitational force having strength proportional to mass content, the postulated tiny particles should be capable of penetrating through all matters having atomic and molecular structures (MAM) to interact with all of their subatomic particles. Further more, to be able to constantly lose energy to MAM to produce gravity and other universal phenomena, the postulated tiny particles must always have higher energy content than the subatomic particles of MAM and there should be high-energy sources to constantly produce them, such as Stars. Since stars also contain MAM having gravity, the postulated tiny particles must be produced by the part of stars under going nuclear reactions only.

It is likely that only one type of tiny particles constantly produced by the nuclear reactions in stars, which has been named neutrinos. Neutrinos however have been found to be too inactive with MAM to produce any universal phenomenon. Since the postulated tiny particles should interact strongly with MAM to produce gravity, light, etc., the author calls the postulated tiny particles NEU to differentiate them from neutrinos. Since the postulated tiny particles should also be produced by the nuclear reactions of stars and, thus, they may either be different from neutrinos or are neutrinos. In the later case, we need to prove that neutrinos interact strongly with MAM to produce gravity and other universal phenomena meaning that the current theory of neutrino is wrong.

Besides inactive neutrinos, the space of the universe is also filled with an atmosphere of NEU (the postulated tiny particles) continuously replenished by all stars producing them at constant rates. Since NEU have the ability to penetrate through MAM, they also fill up the space occupied by all MAM. NEU therefore should be the only possible aether to interact strongly with MAM to produce all universal phenomena, such as gravity, light, charged particles, etc. anywhere in the universe. To prove whether this uniquely important logical conclusion is correct, all universal phenomena must be explainable and predictable by the presence and the unique properties of NEU. Besides, scientific mechanism and evidences must also be found to explain and to prove the interactions of NEU with MAM. Only the correct physics can meet all these difficult scientific challenges, as illustrated in this book.

Mechanical gravity

Scientific force field

Scientifically, force fields do exist. They are however not the matter-free force fields of Faraday but the fields of particles, which produce forces themselves and interact with other matters to produce forces and phenomena. A field of particles alone is a force field because that all particles have kinetic energies capable of producing forces and energies. For example, a field of air particles produces atmospheric pressure on all matters in the field. A particle field alone does not produce a net directional force since all collision forces they produced are in random directions and are cancelled out from one another. The interactions of the field particles with other matters however often result in producing directional forces. For example, the collision interactions of atmospheric NEU with all matters produce gravitational, electric, and magnetic forces. A matter exchanges energy with the field particles to produce directional forces. For example, an electric fan produces wind.

An electric fan interacts with the field of particles of air to produce wind. A constant energy supply is needed to continue to produce wind.

For over a century scientists have accepted the postulation of the existence of matter-free force fields in an attempt to explain universal forces acting from distance but they still have not proven their existence and cannot offer a common-sense-making mechanism for them to mediate (attraction) forces (acting from distance). It will be shown that the real answer is the presence of an atmosphere of NEU to produce universal forces. What a simple scientific answer, which has long been predicted by the teachings of PM! The force fields referred to in this book are therefore particle fields, which only produce on-contact pushing forces.

It will explain that there are special scientific reasons for the atmospheric NEU to produce different universal forces such as gravitational, electrical, and magnetic forces.

NEU – the gravity producing particles

According to the teachings of PM the collision-interactions of matters having mass and relative motions produce all phenomena. Scientists have long predicted the presence of aether, which should be material particles, interacting with ordinary matters to produce gravitational force and light. When two heavenly bodies, Sun and Earth for example, do not contact with each other, they can not interact to produce any force or exchange any energy. Therefore, there must be an atmosphere of material particles to interact with both of them to produce their gravity. The atmosphere of NEU is the only possible material particles to interact with both of them to produce gravity.

According to experimental findings the strength of gravitational force is proportional to the mass contents of MAM. Since all atoms and their subatomic particles such as nuclei and electrons, contain masses, NEU must be able to penetrate through MAM, their atoms and molecules to interact with all their subatomic particles to produce gravitational force without damaging them. NEU therefore must be much smaller even than electrons so that their collisions with subatomic particles are non-destructive to atomic and molecular structures. Besides, NEU must have amazingly dense mass to produce very strong gravitational forces and travel very fast to allow gravitational force to have long-range effect already observed by astronomers. The above required unique properties for NEU have been found to be those of neutrinos suggesting that NEU may be neutrinos. The questions raised therefore are whether NEU are neutrinos and whether neutrinos interact strongly with MAM to produce gravity contrary to general belief? If neutrinos interact strongly with MAM to produce gravity, where is the huge amount of heat these interactions should produce?

Gravity producing mechanism

All stars produce NEU at essentially constant rates of their own, thus, maintaining an atmosphere of NEU having constant concentrations and energy levels at all locations of the universe. Upon entering a matter, such as a planet, NEU continuously collide with its subatomic particles resulting in continuous loss of kinetic energies to them until they finally get out of it. As a result, at equilibrium, the average kinetic energy of NEU inside any matter is lower than those outside and the NEU entering the matter have stronger kinetic energy (faster-moving NEU) than those leaving it. Since at equilibrium, the number of NEU entering a matter should equal to that leaving it, the concentration of NEU inside a matter is higher than in empty space. Also, since the higher-kinetic-energy NEU entering a matter have stronger pushing force than the lower-kinetic-energy NEU coming out of it, there is always an imbalance in their pushing force from NEU surrounding a matter resulting in having a net pushing force towards the matter. For a round massive planet with homogeneous mass distribution there is always a force from the atmospheric NEU, known as gravitational force, pushing all things towards its center.

All matters on Earth are made of atoms and molecules containing nuclei and orbital electrons. Relatively speaking, they are very small particles occupying very large volume of space. Scientists often say that MAM are essentially empty spaces and they believe that neutrinos (NEU) have only rare chance to hit their subatomic particles, thus, can penetrate through MAM essentially freely rarely colliding with them. It is very hard for them to believe that their extremely weak interaction is strong enough to produce gravity and other universal phenomena. However, scientists are changing their minds. Now, they believe that neutrinos do interact with the MAM of Sun before leaving it.

There are unexpected mechanisms that can greatly enhance the collision-interactions between NEU and the subatomic particles of MAM, which are all charged particles. It will explain later that charged particles have very effective way to interact with NEU essentially without producing heat.

No infinitely strong gravity

Without scientific understanding, the mathematical models of the current gravity theories have omitted all the scientific factors affecting the strength of gravitational force besides mass and distance leading to their wrong conclusions, predictions, and mathematical calculations particularly for stars. According to current theories, both the mass contents and the distance between two material objects are the only variables affecting the gravitational force between them. Theory of general relativity further takes into account the shape and the mass distributions of the objects. Still, there are other important scientific factors affecting gravity having been ignored.

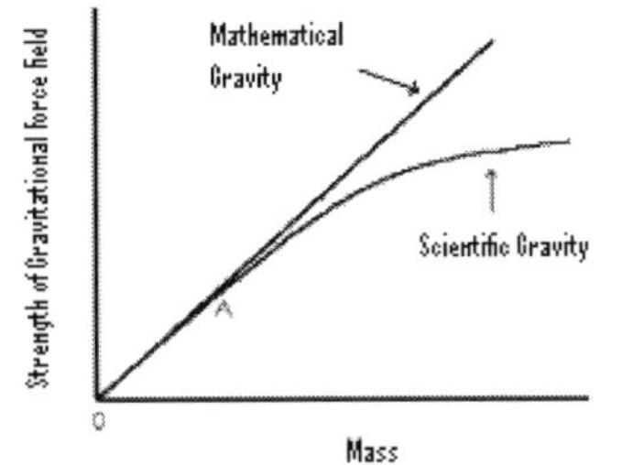

Mechanical gravity has a maximum strength.

Unlike spacetime, an atmosphere of NEU has limited concentration, mass density, and, speed, thus, finite (kinetic) energy. When an object can completely stop NEU inside or around its body, for example, it will experience the maximum pressure or the atmospheric pressure of NEU. It is well known that the atmospheric pressure of air having a maximum pressure on all matters impenetrable by air particles. In a thought experiment that the mass of a planet could continuously increase, its gravity would increase initially proportional to its mass. However, eventually the increase of gravity will begin to slow down deviating from the proportionality, when the atmospheric NEU can no longer reach all the subatomic particles of the planet before their kinetic energy is exhausted. Finally, a maximum gravity will be reached and further increase in mass no longer has effect to the planet's gravity.

For the same reason, the black holes having infinitely strong gravity predicted by the theory of general relativity cannot exist. The nature also cannot have a way to form the so-called "singularity" to produce the so-called big bang.

The gravity of stars

The science of the gravity of stars is much more complex than expected. Take our Sun, a regular star for example, it produces both gravity and antigravity forces not only gravity. The matter at the center of our Sun is antigravity matter producing antigravity since it constantly undergoes nuclear reactions shooting out NEU and other particles. The outer-layer of our Sun is made of MAM, which are gravity matters producing gravity. A baby regular star contains 100% antigravity matter and is therefore an antigravity star but upon aging its antigravity matter continuously converts to MAM to gradually turn into a gravity star. The gravity of a star is therefore not even proportional to its own mass and it varies with aging. Details will be given in the Cosmology chapter.

Space energy

Contrary to general belief that the space of the universe is a vacuum having the lowest energy content, it has higher energy content NEU. NEU are produced by both the radioactive activities and the nuclear reactions of all stars and they have very large kinetic energy. All planets in the universe interact with the atmospheric NEU in such a way that NEU continue to lose energy inside them and, therefore, all planets are the low-energy and high-concentration centers of NEU in the universe. This is apparent since everything around them falls into them.

Gravity stars like our Sun are also low-energy centers of NEU as a whole. However, only their outer layer or the MAM-layer is the low energy centers but their cores are high energy centers having antigravity instead of gravity force.

Mechanical gravity v. spacetime gravity

Einstein's universe has spacetime distortable by environmental matters and their travel speeds. The mass-distorted spacetime is called curved spacetime, which was postulated to produce gravitational force due to the time differences among different locations. In the presence of a massive body spacetime is so distorted that it curves around the body with increasing in distortion curvature upon approaching the body. Inside the massive body the curvature of spacetime reduces upon approaching its center to zero at the center. Curving spacetime causes time to slow down and the difference in time (curvature) between a location closer to and a location farther away from the massive body causes things to fall into the massive body. Curved spacetime according to Einstein is therefore the scientific origin of gravity. Einstein was able to predict some of the gravity-related phenomena with his theory of general relativity, which had shaken the entire world.

Although theory of general relativity postulates no aether to produce gravity, assuming something (spacetime) in space that can be affected (distorted) by environmental matters happens to mimic having-aether science. There therefore exists some similarities between PM and the theory of general relativity as shown in the following picture.

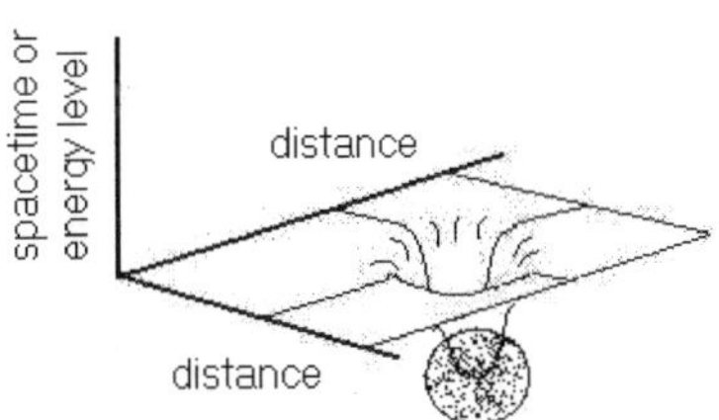

The graph shows the amazing similarity of the variation of the curvature of spacetime to that of the (kinetic) energy of the atmospheric NEU, caused by their interactions with a massive body.

Some amazing similarity exists between the curvature of spacetime and the energy-variation of the atmospheric aether (NEU). Like spacetime, the energy level of the atmospheric NEU remains constant in the absence of a massive body. It also changes around a massive body similar to the distortion of spacetime due to the interactions of the NEU with the massive body. Upon approaching the massive body, the kinetic energy of the atmospheric NEU decreases. Inside the massive body, the kinetic energy of the atmospheric NEU continues to decrease and reaches a minimum level at its center.

The two versions of interpretation of gravity are however incompatible to each other and one of them must be wrong. If theory of relativity were correct, its mathematical model of spacetime should be able to interpret all universal forces such as electric force, the atmospheric pressure of air, etc. but it cannot. Also, if theory of general relativity were correct, the gravity-producing aether (NEU) should never be found.

All no-aether theories were developed due to the longtime failure in finding aether. Many physics books still stated the fact that Einstein's theory of relativity or modern physics had been accepted because some findings had proved that it worked. They further stated that aether did not exist unless theory of relativity was wrong.

Other universal forces

Like the presence of an atmosphere of air, the presence of an atmosphere of NEU should produce a lot more universal phenomena besides gravity. Since NEU are the only particles filling all the space of this universe continuously replenished by stars, they should be responsible for producing all universal phenomena. Besides gravity, scientists have discovered strong nuclear force, weak (nuclear) force, electric force and magnetic force. The question is whether they are all explainable by the presence of an atmosphere of NEU interacting with MAM to produce them? Sure they are as explained throughout the book!

Strong nuclear force

According to modern physics – The nuclei of atoms are made of protons and neutrons (made of protons and electrons). Since protons are positively charged, scientists believe that there exists very strong force to "glue" them together to overcome the repulsion of the positively charged protons to form nuclei and called this universal force the "strong (nuclear) force". They also found this force having very short range of effectiveness unlike the "inverse square force" of the electromagnetic force and gravity, which have very long range. A Japanese physicist Yukawa modeled the strong nuclear force as an "exchange" force in which the exchange particles are pions and other heavy particles. According to standard model of particle physics, the range of a particle exchange force is limited by the uncertainty principle. Protons and neutrons, which make up the nucleus, are themselves made up of quarks, which are held together by the so-called color force.

According to the expanded PM – The atmosphere of air produces atmospheric pressure on almost everything inside its atmosphere. However, the atmosphere of NEU particles does not have an atmospheric pressure on all "matters having atomic and molecular structures" (MAM)   since NEU penetrate right through them. The only thing they may not be able to penetrate through, at least not easily, is the nuclei of atoms, which has amazingly dense mass. There should be a full or partial atmospheric pressure of NEU on atomic nuclei. Unlike gravity, electric, and magnetic forces, strong nuclear force was found to be very short ranged and therefore has the unique properties of an atmospheric pressure, which become ineffective in the presence of very small open space allowing atmospheric NEU to enter or to pass through and therefore should be very short ranged force indeed.

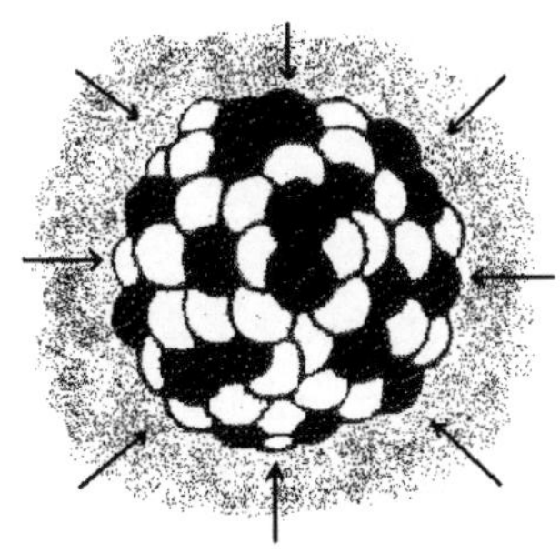

The atmospheric pressure of neutrinos may be the scientific origin of the strong nuclear force.

Strong nuclear force or the atmospheric pressure of NEU should play an important role in the formation of atomic nuclei.

Weak force

Scientists have found the existence of weak (nuclear) force, which causes radioactive decays of some atomic nuclei and instability of cosmic particles.

According to modern physics – According to the standard model the weak interaction involves the exchange of the intermediate vector bosons, the force particles W and Z. Since the mass of these particles is in the order of 80 GeV, the uncertainty principle dictates a range of about 10^{-18} meters, which is about 0.1% of the diameter of a proton. The weak interaction is the only process in which a quark can change to another quark, or a lepton to another lepton – the so-called "flavor changes".

According to the expanded PM – The weak force is an expected universal force simply due to the presence of an energetic dense atmosphere of NEU. The NEU produced by stars are very energetic and their constant bombardments cause cosmic particles and even some atomic nuclei to decay.

Chapter 3 – Charge energies, collision interactions, and light

Current concept of charge energies

It is generally accepted knowledge that there are two kinds of charged particles, the positively charged and the negatively charged. Particles carrying the same charge repulse each other but those having different charges attract each other. Unlike gravitational force acting on everything, electric and magnetic forces are selective forces, which only act on those materials carrying charges and/or having lone-pair orbital electrons.

Ordinary matters MAM do not carry charge. However, they are made of two fundamental charge particles (FCPs): proton and electron. Proton carries positive charge while electron is negatively charged.

Both electric and magnetic forces are explained using the concepts of force-field of Faraday and mathematically by Maxwell's electrodynamics, and modern physics. According to Maxwell, electric force and magnetic force have the same scientific origin giving a combined name: electromagnetic force, which has infinite range obeying the inverse-squire (Columbus) law, thus, having the similar mathematical formulae as for calculating gravitational force. According to the quantum electrodynamics (QED) of modern physics, electromagnetic force is the universal force involving exchanging photons.

The expanded PM

Fundamental charge particles (FCPs)

There are only two stable FCPs, proton and electron, which make up all MAM. It is vitally important to have a correct scientific understanding of these two fundamental particles and their charge properties. A major fraud of the current concept is that it has simply accepted the fact of the existence of FCPs without asking where their charge energy comes from.

The structure of FCPs

The charge phenomena of FCPs (proton and electron) are charge energies, which constantly produce electric and magnetic force. Without charge energy a FCP should be an absolutely uncharged particle (AUP) like a neutrino having much larger sizes and masses. However, being constantly charged, according to the law of conservation of energy an AUP needs constant supply of energy from its surrounding environment to become and maintain charged. Since the atmosphere of NEU is the only matter can reach the AUPs of FCPs even inside MAM, the AUPs of FCPs must obtain their constant charge-energy supply from NEU. The AUPs of both FCPs must have special structural features to effectively and constantly get the (kinetic) energy from the atmospheric NEU of the universe. They may have such structural feature similar to those of windmills and galloping horse lantern to continuously and effectively gain kinetic energy from the NEU of the universe and likely converting it into their own spin energy. By doing so, a FCP becomes a low energy center of the atmosphere of NEU similar to a planet. Like a planet, a FCP should have gravitational force and its own atmosphere of NEU having lower energy but denser concentration of NEU than the atmosphere of NEU of the universe. However, unlike a planet, the AUP of a FCP must have an amazingly dense mass to effectively getting the kinetic energy of NEU by drastically slowing down and even stopping NEU. As a result, at equilibrium the AUPs of FCPs must have an amazingly dense atmosphere of NEU of their own, which should also have a gradient of increasing concentration or decreasing kinetic energy toward their AUPs.

Constantly gaining kinetic energy from NEU the AUPs of both FCPs therefore should have both strong spin and, thus, oscillation energies and their collision interactions further make their atmospheric NEU to spin and oscillate. The atmospheric NEU of a FCP therefore are expected to have both spin and oscillation energies with gradients getting stronger towards their AUPs. The closer the atmospheric NEU of a FCP to the AUP, the stronger their spin and oscillation energy are. This is an important concept since, as it will explain later, both the spin and the oscillation energies of the NEU in the atmosphere of NEU of electrons are their light energies.

A FCP therefore is made of an absolutely uncharged particle (AUP) much bigger than a neutrino particle having an amazingly dense atmosphere of NEU of its own. The concentration of the NEU of the atmosphere of NEU of a FCP should be so high that they can effectively collide interacting with one another and with those NEU entering it to continue to get their kinetic energy.

NEU inside MAM such as Earth are therefore unevenly distributed. There are the free NEU of the universe having higher kinetic energy but the majority of NEU are inside the atmospheres NEU of FCPs. At equilibrium, the same number of NEU enters and comes out of the atmospheres of the NEU of FCPs exchanging energy needed to maintain their charge property or charge energy.

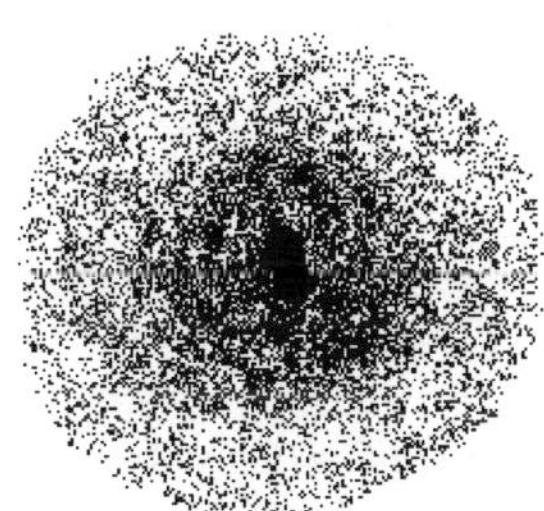

A FCP is made of an AUP having an amazingly dense atmosphere of NEU with a gradient of increasing concentration towards its AUP at center.

The formation of the two FCPs is due to both the huge difference in kinetic energy between their AUP and NEU and that the two AUPs have structural features to effectively and continuously get the kinetic energy of NEU from their collision interactions. Even though we are under an atmosphere of air, air neither produces gravity nor have significantly higher concentration around solid things. This is because that everything around us including air has the same energy level. However, in the presence of a low-energy center air concentration will get higher around it. For example, in the presence of a piece of dry ice, its surrounding air concentration gets higher and more drastically ice will deposit around it.

The charge properties of FCPs

Like a planet, a FCP has gravitational force of its own, which is the atmospheric pressure of the NEU of the universe. Besides, having an extremely dense atmosphere of NEU a FCP should have repulsive force acting on other FCPs to keep them away. These should be the two fundamental forces of a FCP to show its charge property or energy. The size and the unique ability of its AUP to get energy from NEU play important roles in shaping up the charge properties of a FCP. In comparison, although proton is much bigger than electron, both appear to have equal charge power likely due to that the atmosphere of NEU of the universe has a constant energy level. Having much bigger mass and size, protons should have significantly stronger gravitational force than electrons. Meanwhile, electrons should have denser atmosphere of NEU and, thus, relatively stronger repulsion force than protons. The same kind of FCPs (electron-electron or proton-proton) repulses each other due to that having the same size their mutual repulsive forces are maximized and are stronger than their mutual gravitational force. Since a proton is much bigger than an electron, the gravitational force between them becomes the stronger force than their repulsive force resulting in falling into each other initially. However, the repulsive force between them increases rapidly with their approaching to each other. As a result, there is a distance between them where their gravitational force is balanced with their repulsive force and therefore an electron will not fall all the way into a proton.

The above explanation of the charge properties of the two FCPs is applicable and specific only to protons and electrons pairs. For example, it is not applicable to other charge particles found in cosmic ray and produced in high-energy particle accelerators having different AUPs!

Formation of atoms and molecules

An electron falls into a proton by their mutual gravitational force. Meanwhile, their mutual repulsive force increases rapidly upon approaching each other. At equilibrium, the electron oscillates at the distance the two forces are balanced and most likely the electron is orbiting while oscillating like a transverse wave around the proton. This should be the mechanism to form hydrogen and other atoms and molecules. Findings also show that electrons can fall much deeper into a proton in the presence of excess protons to enhance their mutual gravitational forces to form neutrons and atomic nuclei having amazingly dense mass. Neutron was found stable only inside nuclei in the presence of excess amount of protons. Atomic nuclei contain proton(s) and about half its number of electron(s) forming neutrons with protons. The 2:1 ratio allows one electron fall in between two protons maximizing their gravitational interactions.

NEU – FCPs collision interactions

Producing gravitational forces

The gravitational force of MAM acts on all charged and non-charged MAM. It is the differential pushing force between the NEU entering and the NEU leaving a heavenly body. Strong collision interactions between NEU of the universe and MAM must constantly occur to produce and to maintain gravitational force around and inside a heavenly body.

There is a dense energetic NEU atmosphere of the universe, which is even denser inside planets such as Earth. Inside them they collide effectively with the extremely dense atmospheric NEU of all FCPs to produce and maintain their charge properties, atomic and molecular structures. These collision interactions are however among NEU themselves, which do not directly produce heat! Without producing heat, these collision interactions are undetectable leading scientists to mistakenly believe that the neutrinos, which may be NEU, of the universe are very inactive. However, the presence of gravity is itself a strong proof that the collision interactions between the NEU of the universe and those of the FCPs constantly occur resulting in constant losing of their kinetic energy to those of FCPs or MAM producing gravity. The reason is that all FCPs have amazingly dense atmospheric NEU to effectively collide interacting with the NEU of the universe.

Producing both electric and magnetic forces

In the presence of excess of protons at the (+) pole and electrons at the (-) pole of a battery for example the strong interactions between the free NEU of the universe and FCPs are apparent from the long-range electric force they produce, the electric force field, and potential they build between the poles.

It is rather surprising that the fundamental forces of FCPs are far reaching to form the so-called electric and magnetic force fields. Findings show that both fundamental charge forces of FCPs, namely, the repulsive forces and the gravitational forces, are long-range acting spreading throughout the entire electric and magnetic force fields between two poles. In non-charged MAM, although FCPs have strong close interactions with one another to form atoms and molecules, their long-range charge effects apparently have been canceled by one another. Only free FCPs show long-range charge effects to further prove their strong interaction with the NEU of the universe to build their NEU force fields.

Magnets are made of ferromagnetic materials, which are heavy metals and their compounds having such dense-packed crystalline structures that their outermost (valence) electrons interact with neighboring atoms. As a result their atoms or molecules are orderly oriented. A bar or sheet magnet is not charged but it has higher electron density at its (-) pole and higher proton density at its (+) pole. Like an electric field, a magnetic force field is formed between a (+) pole and a (-) pole. Also like an electric force field both fundamental forces of FCPs in a magnetic field extend throughout the entire magnetic force field. The charge power of partially exposed FCPs should not be as powerful as those of free FCPs but in comparison they are much more densely compacted. As a result, magnetic field can be very strong, convenient, economical to use, and to maintain.

Non-charged MAM are compacted FCPs made from their strong charge interactions. The long-range charge effects of the FCPs in these MAM have been cancelled from each other or used up in their internal bonding interactions. This is apparent since their charge properties are essentially undetectable. There is however much weaker charge properties present in all non-charged MAM to express their unique charge interaction properties different from one another, thus, giving them their unique physical and chemical properties. A non-charged particle such as a neutron or a hydrogen particle is not affected by the charge forces inside a magnetic or electric field. However, at close range a polar molecule is affected by charge forces. In water solution a Na^+ ion, for example, attracts many uncharged water molecules to make it hydrated.

Experimental findings have shown that there is significant difference between the "attraction force" and the "repulsion force" of two magnets. The findings show that at longer distances the attraction force between (+) and (-) poles is significantly stronger than their repulsion forces suggesting that the gravitational force of FCPs have longer-range acting than their repulsive force thus giving conventional calculations another deviation. It further means that the gravitational force of a FCP has longer range action than its repulsive force and as a result two (+) or two (-) charge sources should attract each other at long distance, where their repulsive force is weaker than gravitational force.

In a magnetic or electric force field a proton is pushed away by the proton-rich (+) pole with their mutual maximized repulsing force and is pushed towards (-) pole by their mutual maximized gravitational force. For an electron in a magnetic or electric field, it has both maximum repulsing force with the (-) pole and maximum gravitational force with the (+) pole pushing it towards the (+) pole. Please note that the so-called electric and magnetic force fields are both the NEU field. Magnetic force field is utilized to accelerate charge particles in high-energy particle accelerators. A magnetic bar or sheet has a magnetic force field around it. It however does not create a constant flow of NEU and therefore its force field does not have force acting on non-charged and non-ferromagnetic materials. The magnetic force field of NEU is produced by the stationary partially-exposed FCPs at both poles of magnets.

Current concept of positive and negative charge particles have ignored the law of conservation of energy, since it accepts the existence of charge energies without a constant energy-supply source. Besides, having only one fundamental charge force to express their charge properties not two, current concept can not explain why FCPs form atoms and molecules. The expanded PM shows that the atmosphere of NEU of the universe is the common particle field of all universal forces. Their kinetic behaviors allow them to form the two fundamental FCPs with the AUPs of electrons and protons, which then produce both electric and magnetic forces to form atoms and molecules.

Producing the fundamental heating energy of MAM

The collision interactions of NEU of the universe with the FCPs of MAM are mainly those among NEU, which do not directly produce heat. In these interactions the NEU of the universe constantly lose their kinetic energy to FCPs. The energy FCPs constantly getting is likely been converted to the spin and the oscillation energies of their AUPs, which, in turn, constantly transfer their spin and oscillation energies to their own atmospheric NEU. When AUPs oscillate, their FCPs oscillate too. FCPs are therefore spinning and oscillating around their positions in atoms and molecules. Experimental findings do confirm that both electrons and protons are fast rotating and oscillating around their positions. These fundamental oscillation and spin motions of FCPs in MAM should continue to produce heat. The author calls it the fundamental heat of MAM. It is well known that heat propagates through solids, liquids, and gases rather slowly. When the heat continuously produced by the environmental matters is faster than the overall heat-transfer processes, heat energy accumulates by rising temperature or producing more and faster FCP, atomic, and molecular motions until the overall heat-transfer rate catches up with the heat-producing rate to reach an equilibrium temperature. This explains why the centers of planets have the highest equilibrium temperature and why all planets release more heat than what they receive from the Sun's radiation. Having the fundamental heating energy also explains the presence of volcanic activities on Earth and other planets.

Consequences from wrong charge-energy concept

a) Leading to an era of mathematics-based pseudo-physics

Accepting the old concept of charge particles without questioning their energy source has led to the acceptance of Faraday's force-field theory. Essentially, all mathematical theories including Einstein's theories of relativity, their formal-science philosophy believing that universal phenomena can be represented by mathematical equations are based on this force-field theory including Maxwell's theories of electromagnetic wave and electrodynamics. The so-called plenum aether theory is also based on Maxwell's electromagnetic wave equations sharing the same postulations and formal science philosophy with Einstein's theories of relativity. It has been a century since the mainstream of thought based on Einstein's theories of relativity became the officially accepted physics and cosmology.

If we still do not know the existence of air to produce wind, fire, atmospheric pressure, etc. we may have assumed the existence of force fields to derive mathematical equations to represent air-produced phenomena. It is clear that the mathematical approach cannot lead to the discovery of air and scientific understanding of these phenomena! Yet, we have followed Einstein for a century already!

b) Erroneous standard-model particle physics

Unlike the generalized old concept of positive and negative charge particles, the charge properties of FCPs are specific to protons, electrons, and their MAM only. The instruments, such as an instrument to produce magnetic field, made by MAM should not be able to characterize the charge properties of those "charge particles" found in cosmic rays and made in the high-energy accelerators. The knowledge so obtained should be largely wrong particularly when they are very unstable. Therefore, the correctness of the entire standard model particle physics is in doubt.

There is only an atmosphere of NEU in the universe and therefore the AUP of both protons and electrons can only have an atmosphere of NEU of their own to become FCPs making up all MAM. Both FCPs and their MAM cannot have anti-particles and antimatters. It will explain later that a positron should be an electron having high deficiency in NEU content to behave like a "positron" instead of being an anti-particle of electron.

c) Unduly dismissed the gravitational theory of Le Sage

Le Sage in 17th Century has proposed a gravity theory postulating the presence of ultra-mundane corpuscles (tiny particles), having postulated properties surprisingly similar to those of neutrinos (NEU), which took up all space in the universe to interact with ordinary matters to produce gravity. His theory had encountered serious scientific questions unanswerable at the time and had long been abandoned. It is however essentially correct and is compatible with PM. The main problems it had encountered can now be answered as follows.

1) The collisions of postulated particles with MAM must be inelastic in order to produce gravity. According to the calculations of some physicists such as Maxwell, the heat produced by such collisions would vaporize matters in seconds!

This question was raised because of the lack of scientific understanding of the collision interactions between the postulated corpuscles (NEU) and MAM. As explained above the collision interactions to produce gravitational force are those among NEU (the postulated particles) themselves, which do not produce the large amount of heat as expected.

2) According to the calculation of Pierre-Simon Laplace the speed of gravitational force in Le Sage's theory is "at least a hundred millions of times greater than that of light". At lower speeds aberration would cause the apparent force on the orbiting bodies to point slightly behind the current position of the source, causing the bodies to recede from each other. This "aberration" question still bothers many physicists.

The energy variation of the atmospheric NEU around a star or a planet is at equilibrium or steady-state status. Therefore, gravitational force is always there instantly. There is no "speed" of gravitational force and no aberration problem.

3) The presence of an atmosphere of aether should produce frictional force to slow orbiting planets down. Where is the frictional force?

A rotating regular star like our Sun has a fast rotating center where nuclear reactions occur to produce hydrogen and helium, which produces a spiral NEU wind around it. This spiral wind of NEU provides a directional antigravity force to maintain the steady-state planetary orbiting and rotational motions in the NEU wind blowing direction only. Please read Cosmology section for more detailed explanation.

Collisions among FCPs and MAM

Emission of light – a losing NEU process

The correct physics should be consistent and coherent. It is a great pleasure to realize that the fundamental scientific understanding of charge energy and its properties leads to the scientific understanding of all collision interactions, MAM, gravity, and even light. As described earlier, the collision interactions between NEU and FCPs produce gravity but they do not produce the large amount of heat as expected. Instead, they mainly result in the production of both electric and magnetic forces, which is further responsible for the formation of atoms and molecules. Both FCPs are made of their AUPs having an amazingly dense atmosphere of NEU of their own, which can effectively collide interacting with the NEU of the universe.

The second type of collision interactions are among FCPs and MAM, which are those well-known collision interactions result in producing heat and/or light. Experimental findings show that orbital electrons of atoms and molecules jump into lower-energy orbital to emit light. It is reasonable to interpret this process by that to jump into a lower-energy level an orbital electron need to release some NEU to reduce its repulsive force to reach a new balance of charge (gravitational and repulsing) forces with its atomic nucleus. The light produced therefore should be the group of NEU having the same light (spin or oscillation) energy level emitted from the atmosphere of NEU of the orbital electron when it jumps into a lower-energy orbital. Always to emit a group of NEU having the same light-energy level is likely the scientific reason why the light energies produced by the orbital electrons of atoms and molecules are always quantized. Producing light therefore is a NEU-losing or mass-losing process, which, although has not yet been directly detected for physical and chemical reactions, has been confirmed for nuclear reactions.

Lights are therefore the NEU emitted from electrons, which have spin and oscillation energies. As explained earlier that the atmospheric NEU of an electron have spin and oscillation energies with an energy gradient probably covering the entire spectrum of light: the closer to their AUP, the higher their light (spin or oscillation) energy.

FCPs having varying NEU contents

Realizing NEU to be the main component of both FCPs leads to fundamental understanding of the science of their collision interactions. We have found that orbital electrons emit or absorb light upon changing their energy levels. We have further discovered that all collision interactions among FCPs and MAM produce or absorb light including heat energies, since they cause orbital electrons to change energy level. Even weak collision interactions, which only cause molecular motions, involve producing heat and light, such as infrared and microwave lights.

Well-established findings have proven that electrons (and may be protons) in heavier atomic nuclei are very slightly lighter than those in lighter atomic nuclei. These findings suggest that free FCPs may be very slightly heavier than those in light-weight atomic nuclei too. The findings prove that both electrons and protons contain NEU and that their NEU content varies with the environment they are in. Free electrons and probably protons therefore need to release some of their atmospheric NEU to form the nuclei of light-weight atoms and the atomic nuclei in lighter atomic nuclei need to lose even more NEU to form the nuclei of heavier atoms through nuclear reactions. Since in all nuclear reactions the total number of protons and electrons are accounted for, the large amount of energies released in nuclear reactions must come from the large amounts of light-energy-carrying NEU released by the nuclear reactions. The well-established findings of losing weight in nuclear reactions therefore prove that FCPs contain NEU, thus, the proposed structures of FCPs, that a light emission process is a NEU-losing process, and that NEU have mass.

The above findings-based explanations for both nuclear reactions and the origin of light reveal a generalization in explaining all collision interactions among FCPs and MAM including all physical, chemical and nuclear reactions: the energies released or absorbed from these collision interactions come from the emission or the absorption of NEU containing light energies. All these collision interactions involve losing or gaining light-energy-carrying NEU. It has been proven that orbital electrons emit or absorb light-NEU upon switching orbital levels. However, the releasing and the absorbing infrared light are from the collision interactions involving molecular motions, not necessarily changing orbital level.

The mass balance law for both chemical and physical reactions therefore should be modified to take the loss and the gain of the light-energy-carrying NEU into account and it should also be expanded to include nuclear reactions. The scientific community has been wrong by abandoning Mass Balance Law and by postulating that in nuclear reactions mass converts to pure energy, which is not even definable scientifically.

Light – the detectable NEU

Lights are NEU having various levels of spin or oscillation energies. These findings prove that even NEU, the smallest particles of all, are not point or spherical particles since findings show that light (particles) oscillate while traveling showing transverse wave-like properties. However, unlike waves, traveling light particles rarely collide or interfere with one another. The fact that we can see everything very clearly proves that lights of different colors do not interfere with one another. We can even see far away objects such as moon with precise images. The light-(energy-carrying) NEU of remote stars and galaxies coming from millions of light years away are still likely carrying their true images only suffering from redshift. These facts clearly disprove that lights are (electromagnetic) waves. Light-NEU particles interact strongly with MAM to lose their light energy. Even the strongest light, X-Ray, cannot penetrate deep into MAM. All lights are detectable from their strong collision interactions with MAM.

The interactions between light and MAM are so complicate and so hard to understand that it has long become the most controversy scientific topic. For example, the oscillation of light-NEU has led to belief that lights are transverse waves. Further more, the photographs of the scattering light-NEU passing through slit(s) show wave-like patterns appearing to support wave theories of light. Light-NEU particles undoubtedly have particle properties, thus, leading to the wave-particle-duality theory. Findings proved that light and even larger particles such as electrons show wave-like photographic patterns upon passing through single or double slits. To explain these results we need to realize that in atoms and molecules orbital electrons move like transverse waves circulating their atomic nuclei and that orbital electrons are likely responsible for interacting with light NEU particles. Upon passing through narrow slits, Lights or other small particles interact with and reflected by orbital electrons to produce wave-like photographic patterns. These findings therefore prove the expanded PM not that small particles have wave properties.

Undetectable NEU

Light-energies are likely the spin or the oscillation energy NEU carry. Having oscillation property may be the reasons why the light-NEU particles have transverse wave-like properties, why they cannot penetrate through MAM, and why they are detectable. The gravity producing NEU cannot be light-NEU since they should be penetrable through MAM to interact with all subatomic particles of MAM to produce gravitational force proportional to the mass content of MAM. The presence of gravity therefore proves that there is always a very dense atmosphere of undetectable NEU in the universe. Besides, the presence of other universal phenomena such as the presence of FCPs and MAM further proves this fact.

In the night the sky is dark showing that there is no significant amount of light-NEU coming from sky beyond those from stars and moon. On Earth our powerful telescopes are capable of observing stars and galaxies even from millions of light years away often without significant background interference. Although the atmospheric air interferes with the observations of stars and galaxies by X-ray, IR, and radio wave telescopes, most interference can be avoided by putting the telescopes in space, such as Hubble telescopes. The findings further prove the dark-light observation that there is no significant amount of light-NEU arriving at Earth during night hours since we can even detect very weak light radiations such as the cosmic microwave background (CMB), which likely comes from the radiations of the MAM in the intergalactic space surrounding our galaxy, which will be further discussed later. The overall findings suggests that present technology can detect light at all energy levels from X-ray to microwave and besides CMB the only lights coming from all directions at night come directly from stars and moon only.

The dark night phenomena suggest that there are sufficient amount of MAM particles in space to absorb light energies to convert the light-NEU radiated from stars to undetectable NEU. They are so effective in converting light-NEU to undetectable NEU that their overall deflection light-background, which should cover a broad range of temperature, is undetectable!

The presence of gravity and other universal phenomena proves the presence of a dense atmosphere of undetectable NEU easily penetrable through MAM but they interact strongly with MAM to produce many universal phenomena. It is so dense that its NEU can effectively keep air and even hydrogen particles escaping from Earth.

So far nobody believes that lights are neutrino or NEU particles. Scientists do believe that there is a dense atmosphere of neutrinos continuously produced by all stars, which penetrate through MAM such as Earth essentially freely without collision interactions with MAM. They therefore do not believe that neutrinos have anything to do with producing universal phenomena. They still have no reliable method to directly detection neutrinos. The detection method used by the Super-Kamiokande laboratory in Japan assumes that only neutrinos can penetrate deep into Earth to reach the detector – a sensitive light detector. They think that there is only rare chance for neutrinos to collide with MAM particles to occasionally break a chemical bond to produce a faint light. They may have obtained weak signal and detected something. However, little information can be obtained from such a detector. Besides, the detection method is non-specific and it may have detected the existence of rare AUP particles having significantly larger masses than neutrinos, for example.

Although the presence of a dense atmosphere of non-detectable NEU, which may be neutrinos, cannot be directly proven experimentally, its presence is explained and proven by the presence of all universal phenomena. The expanded PM logically concludes from huge amount of findings accumulated over centuries that NEU is the main constituent of FCPs and the free NEU of the universe interact strongly with FCPs and their MAM to produce all universal phenomena. Since the atmospheric NEU of the universe are the only possible thing to produce universal phenomena anywhere in the universe, all universal forces such as gravitational, electric and magnetic forces, and light, must be produced by them. Light happens to be the directly detectable NEU. Stars constantly produce light-NEU, which are constantly converted to undetectable NEU by MAM to maintain a very dense atmosphere of undetectable NEU in the universe. Capable of offering a simple logical consistent interpretation of the universe including all its universal phenomena, the presence of a dense atmosphere of undetectable NEU should be considered proven.

Light – historic background

A historic review on light theories is give below based on Wikipedia, the encyclopedia on Internet which typically reflects the views of the mainstream of thought.

Particle theory

Ibn al-Haythan (Alhasen, 965-1040) proposed a particle theory of light in his Book of Optics (1021). He held light rays to be streams of minute energy particles that travel in straight lines at a finite speed. He states in his optics that "the smallest parts of light," as he calls them, "retain only properties that can be treated by geometry and verified by experiment; they lack all sensible qualities except energy." Avicenna (980 – 1037) also proposed that "the perception of light is due to the emission of some sort of particles by a luminous source".

Pierre Gassendi (1592-1655), an atomist, proposed a particle theory of light which was published posthumously in the 1660s. Isaac Newton studied Gassendi's work at an early age, and preferred his view to Descartes' theory of the plenum. He stated in his Hypothesis of Light of 1675 that light was composed of corpuscles (particles of matter) which were emitted in all directions from a source. One of Newton's arguments against the wave nature of light was that waves were known to bend around obstacles, while light travelled only in straight lines. He did, however, explain the phenomenon of the diffraction of light (which had been observed by Francesco Grimaldi) by allowing that a light particle could create a localised wave in the aether.

Newton's theory could be used to predict the reflection of light, but could only explain refraction by incorrectly assuming that light accelerated upon entering a denser medium because the gravitational pull was greater. Newton published the final version of his theory in his Opticks of 1704. His reputation helped the particle theory of light to hold sway during the 18th century. The particle theory of light led Laplace to argue that a body could be so massive that light could not escape from it. In other words it would become what is now called a black hole. Laplace withdrew his suggestion when the wave theory of light was firmly established.

Wave theory

In the 1660s, Robert Hooke published a wave theory of light. Christiaan Huygens worked out his own wave theory of light in 1678, and published it in his Treatise on Light in 1690. He proposed that light was emitted in all directions as a series of waves in a medium called the Luminiferous ether. As waves are not affected by gravity, it was assumed that they slowed down upon entering a denser medium. Thomas Young's sketch of the two-slit experiment showing the diffraction of light. Young's experiments supported the theory that light consists of waves.

The wave theory predicted that light waves could interfere with each other like sound waves (as noted around 1800 by Thomas Young), and that light could be polarized. Young showed by means of a diffraction experiment that light behaved as waves. He also proposed that different colors were caused by different wavelengths of light, and explained color vision in terms of three-colored receptors in the eye.

Another supporter of the wave theory was Leonhard Euler. He argued in Nova theoria lucis et colorum (1746) that diffraction could more easily be explained by a wave theory. Later, Augustin Jean Fresnel independently worked out his own wave theory of light, and presented it to the Academie de Science in 1817. Simeon Denis Poisson added to Fresnel's mathematical work to produce a convincing argument in favour of the wave theory, helping to overturn Newton's corpuscular theory.

The weakness of the wave theory was that light waves, like sound waves, would need a medium for transmission. A hypothetical substance called the luminiferous aether was proposed, but its existence was cast into strong doubt in the late nineteenth century by the Michelson-Morley experiment.

Newton's corpuscular theory implied that light would travel faster in a denser medium, while the wave theory of Huygens and others implied the opposite. At that time, the speed of light could not be measured accurately enough to decide which theory was correct. The first to make a sufficiently accurate measurement was Leon Foucault, in 1850. His result supported the wave theory, and the classical particle theory was finally abandoned.

Electromagnetic wave theory

Faraday (1832) developed the mathematical concept of the 'electro-magnetic force field' as a way of mathematically describing action-at-a-distance for charged particles (i.e. electrons and protons). This is a continuous mathematical 'plotting' of the effects (forces and thus accelerated motions) that matter has on other matter in the Space around it, thus it is a description of effects rather than causes.

Based on Faraday's force field concept and assuming the existence of both electric and magnetic force fields, Maxwell derived electromagnetic wave equation. According to this equation, the self-propagating electromagnetic waves would travel through space at a constant speed, which happened to be equal to the previously measured speed of light. From this, Maxwell concluded that light was a form of electromagnetic radiation: he first stated this result in 1862 in On Physical Lines of Force. In 1873, he published A Treatise on Electricity and Magnetism, which contained a full mathematical description of the behaviour of electric and magnetic fields, still known as Maxwell's equations. Soon after, Heirich Hertz (1888) discovered radio waves, a invisible light produced by the oscillation of electrons. This experimental finding has been used to confirm Maxwell's theory that light is electromagnetic wave.

There were two unscintific conclusions leading to the mess of the physics and cosmology over the past one and half century. First of all, despite the existence of force fields is true or false, a phenomena cannot be one or a set of equations since they are produced by certain material particles with different scientific reasons. For example, a fire can be produced by many matters for many different reasons, which cannot be any mathematical equation. Therefore light and gravity cannot be equations derived mathematically. Secondly, finding radio wave does suggest that light is produced by electrons but it does neither prove the existence of force fields nor prove that radio wave is electromagnetic wave.

The postulation of the existence of both electric and magnetic force fields in space without an energy source to constantly maintain them violates the law of conservation of energy and thus PM. Both Einstein's (no-aether) theory of relativity and plenum-aether theory are the twins born from the Maxwell equations. They both therefore share the same postulations such as force fields and invariance of light speed, orthodoxies, and both are formal sciences. They are however contradictory to each other with one a no-aether theory and the other a having-aether theory.

The special theory of relativity

The wave theory was considered wildly successful in explaining nearly all optical and electromagnetic phenomena. By the late nineteenth century, however, a handful of experimental anomalies remained in direct conflict with the wave theory. One of them involved a controversy over the speed of light. The constant speed of light predicted by Maxwell's equations, which was considered confirmed by Michelson-Morley experiment contradicted the mechanical laws of motion that had been unchallenged since the time of Galileo, which stated that all speeds were relative to the speed of the observer. In 1905, Albert Einstein revised the Galilean model of space and time to account for the constancy of the speed of light. Einstein formulated his ideas in his special theory of relativity, which radically altered humankind's understanding of space and time.

Michelson-Morley experiments are based on the postulation that lights are electromagnetic waves interfering one another in space. Since the postulation is wrong, this experiment should not work and should obtain negative result as they have been and it does neither prove the invariance of light speed nor the absence of aether.

Assuming that light has an absolute speed certainly has violated the fundamental law of PM that teaches that relative speed exists among all matters. Changing time and space from absolute to variables of mass and speed, Einstein has created a fantasy universe of four-dimensional spacetime having matter-free energies, antimatters, capable of time travel to the past and future, etc. It is a mathematical game with attractive financial reward, which many physicists and mathematicians love to play.

Quantum theory

Another anomaly that arose in the late 19th century involved a contradiction between the wave theory of light and measurements of the electromagnetic spectrum emitted by thermal radiators, or so-called black bodies. Physicists struggled with this problem, which later became known as the ultraviolet catastrophe, unsuccessfully for many years. In 1900, Max Planck developed a new theory of black-body radiation that explained the observed spectrum correctly. Planck's theory was based on the idea that black bodies emit light (electromagnetic radiations) only as discrete bundles or packets of energy. These packets were called quanta, and the particle of light was given the name photon, to correspond with other particles being described around this time, such as the electron and proton.

Quantum theory has been expanded to suggest that in microenvironmets, all energies are quantized. It is however even contradictory with Einstein's general theory of of relativity since gravitational force is not quatized. The author suggested that the quantum effect of light is due to that the unique structure of electrons and an orbital electron always emits a group of light-NEU having the same energy when it jumps into a lower energy orbital.

The wave–particle duality theory

Another experimental anomaly was the photoelectric effect, by which light striking a metal surface ejected electrons from the surface, causing an electric current to flow across an applied voltage. Experimental measurements demonstrated that the energy of individual ejected electrons was proportional to the frequency, rather than the intensity, of the light. Furthermore, below a certain minimum frequency, which depended on the particular metal, no current would flow regardless of the intensity. These observations clearly contradicted the wave theory. In 1905, Einstein resurrected the particle theory of light to explain the observed effect forming the basis for wave–particle duality and much of quantum mechanics.

The modern theory that explains the nature of light includes the notion of wave–particle duality. The theory states that everything has both a particle nature and a wave nature, and various experiments can be done to bring out one or the other. The particle nature is more easily discerned if an object has a large mass, and it was not until a bold proposition by Louis de Broglie in 1924 that the scientific community realized that electrons also exhibited wave–particle duality. The wave nature of electrons was experimentally demonstrated by Davission and Germer in 1927.

Speed of light

The physics of light, the detectable NEU, should not be different from any other material particles. However, light speed has been the center of scientific controversy in the past century. It has been over a century since light has been taught to be electromagnetic wave, an equation derived by Maxwell based on Faraday's force-field theory. According to the equation light speed should be a constant, which has led to Einstein's postulation that light travels at an absolute constant speed in space! Having absolute constant speed means that light has no relative motion with all other matters, thus, violating the fundamental law of PM that all matters have relative motions. Even today, physicists still rely on the negative findings of the over-a-century-old Michelson-Morley experiment to prove this postulation. This experiment is based on the reasoning that since Earth has both rotational and orbital motions light on Earth coming and going in different directions should travel at slightly different velocities. It used interferometer to detect this expected micro-differences of light speed with the postulation that lights are waves interfering with one another. The experiment split the same light beam into two traveling at directions and then recombined them. Physicists believe that having different velocity would have changed their wavelengths and the recombined light beams should have slightly different wave lengths and should interfere with each other producing changed interference photographic patterns. The experiment however detected no such light-interference pattern change and the generally accepted conclusion was that there was no aether in space to transmit light and that the null finding supported Einstein's postulation that light traveled at an absolute constant speed in space without a medium (aether) to transmit it. Now the (electromagnetic) wave theory of light has been proven wrong. And Light-NEU particles in space essentially do not collide with one another to produce interference. Therefore, interferometer should not have detected any light-interference pattern change in the experiment! Michelson-Morley experiment was an experiment used invalid detection method.

In 1992 NASA's COBE satellite mapping study of the cosmic microwave background (CMB) found by A. Penzia and R. Wilson in 1963 happened to be based on the same scientific principle of the Michelson-Morley experiment. This experiment directly determined CMB light energy coming from different direction. This time the NASA scientists intended to prove that the CMB light coming from all directions have the same energy level indicating the same light speed. Besides, physicists were very interested in this study since having a homogeneous residue light background leftover is predicted by the big bang theory and they need experimental findings to prove this theory. However, this study surprisingly discovered dipole anisotropy property of the CMB. What NASA had found was that the energy of the CMB light coming from the southwest direction was the highest (darker color) indicating the light velocity is the fastest. Meanwhile, the energy of CMB coming from the northeast direction is the lowest (lighter color) or the light velocity is the slowest.

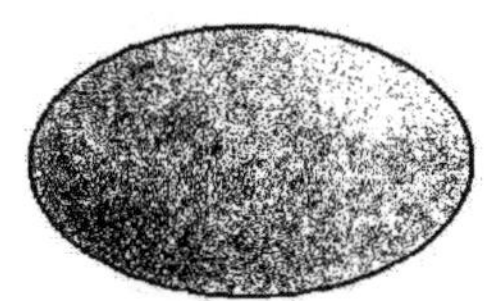

CMB mapping study by NASA
discovered its anisotropy property.

The change in energy level is gradual from the southwest to northeast making this dipole anisotropy property very difficult to be explained by any other scientific reason. NASA therefore had found that light does travel at different speed. In other words, NASA has disproved the fundamental postulation of Einstein's theories of relativity based on that light travels at an absolute constant speed in space! However this important conclusion had essentially been ignored by the scientific community! Instead, an artificial correction was made to the dipole anisotropy of CMB to make the CMB mapping result homogeneous from all directions, which has been advertised as the experimental proof of the big bang theory! What a joke on science! A few physicists however have discussed the above findings and their comments are given below.

From the "Little Book of the Big Bang" by Craig J. Hogan published in 1998: "Maps of the cosmic background radiation tell our velocity through the universe quite precisely – both how fast we are going and in what direction – to an accuracy of a few percent, better than a car speedometer. It is significant that there is such a thing as a velocity through the universe, because absolute velocity is not defined by the laws of (modern) physics per se, which recognizes only relative velocity; indeed, this "principle of relativity" is a cornerstone of our theory of spacetime. However, the distribution of the matter and radiation of the universe does define a "preferred frame" that, at each position, in the frame in which the universe looks the same in all directions. The solar system is moving relative to that frame, with a velocity of 375 kilometers per second, toward the head of the constellation Virgo." Also, in the book "Ripples in the Cosmo" by Michael Rowan-Robinson published in 1993: "When allowance was made for the Earth's motion around the Sun and the Sun's motion around the galaxy, the net result was that our Galaxy was moving through space at a speed of 600 kilometers per second. This seemed an alarmingly high speed. There is a curious aspect to this. The famous Michelson-Morley experiment of 1885 showed that the speed of light in a vacuum is the same whatever the speed of the observer. This destroyed the concept of the aether, which had dominated nineteenth century physics, and launched the way for the special theory of relativity in which all observers in uniform motion with respect to each other are equivalent. In cosmology, however, the situation is different. There appears to be an absolute cosmological frame of rest in which the universe looks perfectly isotropic. We are not in that frame so we see the microwave background dipole. In writing about the discovery of the dipole for *nature* in 1977, I ironically headlined my piece "aether drift detected at last."

Creating an "absolute or preferred cosmological frame of rest simply cannot scientifically explain the anisotropy findings of CMB. Both physicists have suggested that Earth is moving against a strong aether wind at ~375-600 kilometers/second. It is doubtful that encountering such a strong aether wind is essentially undetectable. It is more likely that our galaxy as a whole moves against its surrounding environment where the CMB come from. This possibility will be further discussed later.

Other findings having often been used to prove that light travels at an absolute constant speed are those astronomical findings of binary star systems with their planes seen sideways from us. Each star alternately moves toward and then away from us as it orbits around its companion star. These movements should result in sending us alternative non-shift, redshift, non-shift, and blueshift spectra. If these spectra of lights travel at the same speed, Astronomers should always get the spectra from all of them having the above unscrambled sequence. Although astronomers have found some unscrambled spectra from some binary star systems, only long term study of the spectra sequence of many binary stars can determine whether light speeds are different or not.

The light-speed controversy has been largely from the extremely high speed of light difficult to measure accurately. It is a really miserable fact that the correctness of the entire physics and cosmology already accepted worldwide for a century already has to be betted on whether the light speed in space is absolute or not!

Universal redshift of star lights

Doppler Effect of sound

The Doppler effect of sound was first investigated and explained by an Austrian physicist C. J. Doppler in 1842. He studied auditory phenomena, for example, the sound of a police car's siren grew higher in pitch as the car approached but lower as it drove away. He found that if there was relative motion between the sound source and the receiver, the frequency of the sound was altered: the frequency of sounds were raised if the source and the observer were moving toward each other and lowered if they were moving apart. It is a medium effect detectable by measuring the changes of wavelength, frequency and/or the relative velocity of the so-called mechanical wave of sounds.

It is also known that change in energy and/or concentration level of transmission medium of sound also produces Doppler Effect-like phenomena without the presence of any relative motion. For example, both the temperature and the concentration of air are often lower on the top of the mountain than those at bottom. When a sound travels from the bottom of the mountain to the top, its tune gets lower thus getting an effect mimicking red-shift Doppler Effect in the absence of relative motion. Besides, the sound gets weaker too, letting observer to believe that the source of the sound is farther away than it really is.

Doppler Effect of light

Lights are emitted by orbital electrons and are individual light-NEU particles traveling through space to reach observers. Unlike sound, light-NEU particles travel through space not in concentration-wave form and without a medium. It means that lights should have relative-motion Doppler Effects but no "medium" Doppler Effect. Light therefore should have relative motion caused redshift and blueshift. The speed of sound is directly affected by the energy-change of the medium (air). For example, sound travels slower in winter and faster in summer. However, since there is only the atmosphere of NEU in the space of the universe, the light-NEU particles should travel in real vacuum without medium effect.

There are both energy and concentration variations of the atmosphere of NEU in the universe. For example, Planets and their surroundings are low energy but high concentration centers of the atmospheric NEU. Besides, the energy and the concentration levels of NEU inside galaxies should be higher than those in the starless intergalactic spaces (IS). However, the variations of both energy and concentration of the atmospheric NEU should not affect the speed and the energy of the light-NEU, which travel through space to reach us without collision interacting with other NEU or MAM. Those light-NEU blocked, scattered by MAM or collided with NEU should not reach us and should not give us the true images, e.g., of stars.

Universal redshift and big bang theory

In 1929 Hubble discovered redshifts in the lights of all remote stars and galaxies and that the size of the redshifts was found to be proportional to their distances from Earth. This relationship is the so-called Hubble Law.

Hubble Law plot: The size of the redshift of a star or galaxy light is proportional to its distance from Earth.

The meaning and the importance of Hubble's findings to modern physics and modern cosmology is summarized by David Wands:

"The observational data available to Hubble by 1929 was sketchy, but whether guided by inspired instinct or outrageous good fortune, he correctly divined a straight line fit between the data points showing the redshift was proportional to the distance. Since then much improved data has shown the conclusion to be a sound one. Galaxies are receding from us, and one another, as the Universe expands. Within General Relativity, the theory of gravity proposed by Albert Einstein in 1915, the inescapable conclusion was that all the galaxies, and the whole Universe, had originated in a Big Bang, thousands of millions of years in the past. And so the modern science of cosmology was born."

The big bang theory is based on Hubble's findings and Hubble Law, which were interpreted by that the universe is uniformly expanding based on the following logic: If we go back in time, the universe would get smaller, denser, and hotter. Eventually, if it does not become zero or nothing, it may become very small. At that time, it should be infinitely hot, dense, and have infinitely strong gravity. Big bang theory calls it "singularity" and postulated that the universe started with a big explosion of a singularity followed by expansion. According to modern cosmology the slope of the Hubble Law plot offers an estimation of the age of the universe – about 15 billion years since the big bang.

Mini-aether theory

Findings have confirmed besides the presence of universal redshift of star lights, the gravitational redshift of light, and the bending of star lights by gravitational force. Without medium effect, both universal redshift of star lights and Hubble Law should lead to the conclusion that the universe is uniformly expanding. However, the relative-motion caused Doppler Effect cannot explain gravitational redshift of light, which appears to be a much stronger effect than the universal redshift, which takes many light years of distances to develop. Besides, the smooth bending of light by gravitational force also remains unexplainable.

To explain all the above phenomena the author proposes a mini-aether theory postulating the presence of an atmosphere of mini-aether particles having particle sizes much smaller than NEU.

Again the constant existence of the atmosphere of mini-aether suggests that they are also produced by the nuclear reactions in stars to constantly replenish them like NEU. It is therefore possible that a single particle of NEU has an atmosphere of mini-aether, thus, having charge energy. If so, individual particle of NEU should have a repulsive force on other NEU keeping all NEU from direct contact with one another. Mini-aether particles are therefore also part of the compositions of the two FCPs. A FCP therefore has an atmosphere of NEU and each NEU has an atmosphere of mini-aether. In the presence of a dense atmosphere of mini-aether, regular stars and planets should have gravitational force on NEU particles including light. This gravitational force should produce gravitational redshift of light and it should also bend the light. Besides, the effect of the universal redshift of star lights is also expected from the presence of an atmosphere of mini-aether in the absence uniform expansion of the universe. Due to that they are much smaller than NEU, mini-aether particles should have even much stronger interactions with MAM and as a result their concentration in space is much lower than that inside heavenly bodies. Consequently, the effect of the universal redshift of star lights is much weaker than the effect gravitational redshift of light.

Mini-aether therefore should be part of the composition of FCPs and participate in all nuclear, chemical and physical reactions exchanging energies with NEU and MAM. The drag effect of the atmospheric mini-aether should gradually slow down both the traveling speed and the spin speed of light-NEU in space to produce the universal redshift effect. In addition, there is a very high concentration of atmospheric mini-aether inside and around regular stars and planets to produce gravitational light-bending and redshift effects. Unfortunately even NEU are too small to be directly detectable now and this mini-aether theory will likely remain hard to prove for sometime.

If the universe is expanding, the atmospheric NEU constantly produced by all galaxies should be the main force to push galaxies away from one another. However, in the inner part of the universe galaxies should experience balanced forces from all directions. Only the outermost galaxies should experience a force by the atmospheric NEU pushing them away from the universe. It means that universe should not be uniformly expanding. Astronomers have confirmed that the most remote galaxies are receding at accelerating velocity, thus, having proved this reasoning.

It is reasonable to assume that the presence of mini-aether is the main cause of the universal redshift for the inner part of the universe but the expansion of the universe also contributes to the redshift of star light for the outer part of the universe.

Cosmic microwave interference background (CMB)

CMB has the characteristics of black body radiation proving that it is radiated by MAM at very uniformly low temperature (~2.7 K). Having billions of stars distributed inside our and other galaxies, planets and asteroids inside all galaxies unlikely can have such a uniformly low temperature. Only the matters inside an intergalactic space (IS) region, such as the IS surrounding our galaxy, may have such a uniformly low temperature and is likely the source of CMB radiation. All IS are likely the dumping ground for surrounding galaxies and may have high content of dead stars, planets, and asteroids.

Today's theoretical physicists have been excited by the discovery of CMB and believe that it is the residue radiation predicted by the big bang theory. They expected a homogenous distribution of the CMB but in 1992 the study of NASA found anisotropy property of CMB. Some physicists suggested that finding anisotropy property in CMB meant the discovery of aether and that Earth or the entire galaxy was moving in aether at a very high absolute velocity, around 600 kilometers per second.

The suggestion that Earth constantly encounters a strong aether (NEU) wind does not agree with all findings. For example, in the presence of such a strong NEU wind Earth's gravity should be unevenly distributed and should change periodically daily but this is not true. Besides, in the presence of strong NEU wind all the light sources on Earth should have anisotropy property having the mirror image of the CMB background, which has not yet been discovered. There are therefore strong scientific evidences against the suggestion of the presence of a strong NEU (aether) wind.

Moving against a strong wind of NEU, all the light sources on Earth should have an anisotropy property having the mirror image of CMB.

It is also possible that instead of moving against "aether", our galaxy moves with the NEU wind. Since CMB is produced by MAM having very uniformly low temperature, ~2.7 K, the CMB producing matters cannot be inside our galaxy or any other galaxy where billions of stars live. Since CBM comes from all directions and it has anisotropy property, the findings suggest that our galaxy as a whole moves inside an environment that radiates CMB. The environment our galaxy is moving in has to be the vest starless IS surrounding our galaxy and it should have many dead stars, planets, and asteroids. Without alive stars, all matters inside the IS should have uniformly low temperature!

The remaining question is why CMB is smooth instead of granular like the telescopic findings of stars? The following reasons may have contributed to the smoothness of CMB. The extremely low temperature in the IS (~2.7 K) may have shattered all crystalline structures of MAM causing dead stars and planets to be easily breaking up. It is likely that the IS is a very windy place since there are many NEU winds from surrounding galaxies constantly blowing into it. Strong NEU winds therefore may also contribute to frequent collisions breaking up dead planets and stars there. Besides, experimental method to measure CMB may have contributed to the smoothness of the overall CMB signal, since telescope was not used. The measurement of average weak energy in an area should smooth the CMB signal out. The breaking up of many dead stars and planets would greatly enhance both their CMB radiation and the smoothness of their CMB signal.

Chapter 4 – Cosmology

Modern cosmology

Before the discovery of expanded PM, PM essentially has nothing to offer on the scientific topics of universal phenomena and cosmology beyond its old prediction that there must be material particles (aether) capable of taking up all space in the universe to produce light and universal forces. The presently accepted modern cosmology is the (no-aether) universal views of modern physics based mainly on theory of relativity and big bang theory. Now the expanded PM offers both scientific interpretation of all universal phenomena and a new cosmology having a scope already broader than all mathematical theories combined. Unfortunately, it is fundamentally incompatible with all mathematical theories including modern physics and modern cosmology, which have long been accepted and supported by the scientific communities, their authorities and taught in schools and colleges.

Modern cosmology is mainly based on the following postulations and beliefs, 1. There is nothing (no aether) in the universe to produce and to transmit light and universal forces. 2. Light speed is absolute but time and dimension are not. 3. All universal phenomena should be interpreted by a mathematical model (theory) including the universe. 4. The universe is uniformly expanding started from a big bang of a "singularity" having infinitely strong gravity, density, and high temperature. 5. Stars are formed from gravitational pull among hydrogen and helium molecules. 6. In the mathematical model of modern cosmology gravitational force is solely responsible for forming stars, starting, and maintaining nuclear reactions at the center of stars, and crashing atoms and molecules to form (dead) dense stars.

Both modern cosmology and astronomy are also based on very large amount of astronomical observatory findings, knowledge of physics, chemistry, and biology. It is useful to summarize and review modern cosmology and astronomy before introducing the new cosmology of the expanded PM. Modern cosmology and astronomy given here can be found in basic cosmology text books and from Internet scientific educational websites and is therefore considered proven, well established, and accepted by the scientific community.

Star formation

According to modern cosmology the universe began about 15 billion years ago from a "big bang", a cosmic explosion of a singularity that resulted in an expanding cloud of the two lightest elements, hydrogen and helium. Where there were higher concentrations of gases, the mutual gravitational attractions of the gas molecules led to the growth of the first generation of stars! As more and more material fell into a new star, the pressure at its center finally became high enough to start the process of nuclear fusion, in which the nuclei of hydrogen and helium merge to form heavier elements. This was accompanied by the release of energy made stars shining. Eventually, all the hydrogen and helium, and those products that could be used to generate energy in the core of the star, were exhausted, and the nuclear furnace was extinguished. As a result, the outer layers of the star could no longer resist the central force of gravity, which was pulling the star's outer matter inward toward its core. What happened next depended on the mass of the star. Very large stars become supernovas, exploding violently spewing much of the stellar material into space and their very strong gravity then crash the atoms and molecules of these stars into very dense fused mass known as black holes and neutron stars. For small stars the process was slower; instead of an explosion, elements from the star's interior zones rose to the surface and were then lost to space when the outer layers blew off, again, leaving strong gravity compressed dead stars known to be white and brown dwarfs.

Hubble's findings of universal redshift of star and galaxy lights and Hubble Law have led to the development of big bang theory, on which the modern cosmology is based. It is very hard to believe that gravitational force among dispersed hydrogen and helium gases is strong enough to collect these gases to form stars. Even if stars are formed from gravitational pull and stars have enough hydrogen and helium fuel to undergo nuclear reactions for billions of years, cosmologists still cannot explain how stars can continue to extract pure hydrogen, bring it to their centers, and remove nuclear wastes from its center to maintain essentially constant nuclear reaction rates for billions of years.

A (gravitational) pressure does not automatically and constantly produce heat. Based on our knowledge, it may cause chemical and physical changes, which either release or absorb heat. These changes only occur over a very short period of astronomical time and the heat energies they produced are too small and too slow to ignite nuclear reactions. If gravitational pressure provides heat to ignite and maintain nuclear reactions, which should occur at a quarter distances inside stars, where their gravitational forces have reached maximum, not at their centers, where no gravitational force exists!

Based on astronomical findings modern cosmology teaches that large and medium size stars explode to end their lives upon using up their hydrogen and helium fuels for their nuclear reactions, which keep strong gravitational pull from collapsing stars. If it were true, the cooling-off or dying stars should collapse to end their lives without exploding up first. If it were true that neutron stars and white dwarfs were dead stars, they should not continue to release strong energies.

Neutron Stars

Neutron stars are about 20 km in diameter and have the mass of about 1.4 times that of our Sun. They are so dense that one teaspoonful would weigh a billion tons on Earth! Because of its small size and high density, a neutron star possesses a surface gravitational field about 2×10^{11} times that of Earth. Neutron stars also have magnetic fields a million times stronger than the strongest magnetic fields produced on Earth.

Neutron stars are one of the possible ends for a star. They result from massive stars, which have mass greater than 4 to 8 times that of our sun. After these stars have finished burning their nuclear fuel, they undergo a supernova explosion. This explosion blows off the outer layers of a star into a beautiful supernova remnant. The central region of the star collapses under gravity. It collapses so much that protons and electrons combine to form neutrons, hence the name "neutron star".

Neutron stars may appear in supernova remnants, as isolated objects, or in binary systems. One neutron star is thought to have planets. When a neutron star is in a binary system, astronomers are able to measure its mass. From a number of such binaries seen with radio or X-ray telescopes, the neutron stars were found to have masses of about 1.4 times that of the Sun. For binary systems containing an unknown object, this information helps distinguish whether the object is a neutron star or a black hole, since black holes are more massive than neutron stars.

Pulsar – Pulsars are rotating neutron stars. And pulsars pulse because they rotate! Pulsars were first discovered in late 1967 by graduate student Jocelyn Bell Burnell as radio sources that blink on and off at a constant frequency. Now we observe the brightest ones at almost every wavelength of light. Pulsars are spinning neutron stars that have jets of particles moving almost at the speed of light streaming out above their magnetic poles. These jets produce very powerful beams of light. For a similar reason that "true north" and "magnetic north" are different on Earth, the magnetic and rotational axes of a pulsar are also misaligned. Therefore, the beams of light from the jets sweep around as the pulsar rotates, just as the spotlight in a lighthouse does. Like a ship in the ocean that sees only regular flashes of light, we see pulsars turn on and off as the beam sweeps over the Earth. Neutron stars for which we see such pulses are called "pulsars", or sometimes "spin-powered pulsars," indicating that the source of energy is the rotation of the neutron star. Some pulsars also emit X-rays.

White Dwarf Stars

White dwarf is produced when a low to medium mass star dies. These stars are not heavy enough to generate the core temperatures required to fuse carbon, and after they have become a red giant during their helium-burning phase, they will shed their outer layers to form a planetary nebula, leaving behind an inert core consisting mostly of carbon and oxygen.

This core has no further source of energy, and so will gradually radiate away its energy and cool down. The core, no longer supported against gravitational collapse by fusion reactions, becomes extremely dense, with a typical mass of about half that of the sun contained in a volume about equal to that of the Earth. Many white dwarfs are approximately the size of the Earth, typically 100 times smaller than the Sun. They may have the same mass as the Sun and so are very compact. A radius, which is 100 times smaller, implies that the same amount of matter is packed in a volume that is typically $100^3=1,000,000$ smaller than the Sun and so the average density of matter in white dwarfs is 1,000,000 times denser than the average density of the Sun.

Eventually, over hundreds of billions of years, white dwarfs cool to temperatures at which they are no longer visible. However, over the universe's lifetime to the present (about 15 billion years) even the oldest white dwarfs still radiate temperatures of a few thousand Kelvin's.

White dwarf stars are extremely hot; hence the bright white light they emit. This heat is a remnant of that generated from the star's collapse, and is not being replenished (unless it accretes matter from other nearby stars). However, since white dwarfs have an extremely small surface area from which to radiate this heat energy, they remain hot for a long period of time. Eventually, a white dwarf will cool into a black dwarf. Black dwarfs are ambient temperature entities and radiate weakly in the radio spectrum, according to theory. However, the universe has not existed long enough for any white dwarfs to have cooled down this far yet, so no black dwarfs are thought to exist. Many nearby, young white dwarfs have been detected as sources of soft X-rays (i.e. lower-energy X-rays); soft X-ray and extreme ultraviolet observations enable astronomers to study the composition and structure of the thin atmospheres of these stars.

White dwarfs cannot be over 1.4 solar masses, the Chandrasekhar limit, but there is a working method to get them over this limit. A white dwarf can accrete material from a companion. Unlike a nova, the material accretes slowly and remains stable. The mass of the white dwarf increases until it hits the 1.4 solar mass limit at which degeneracy pressure cannot support the star. This creates a type Ia supernova and is the most powerful of all the supernovae.

White dwarf was discovered in 1862 by Alvan Graham Clark by discovering a dark companion of the brightest star Sirius (Alpha

Canis Majoris). The companion, called Sirius B or the Pup, had a surface temperature of about 25,000 K, so it was classified as a hot star. However, Sirius B was about 10,000 times fainter than the primary, Sirius A. Since it was very bright per unit of surface area, the Pup had to be much smaller than Sirius A, with roughly the diameter of the Earth.

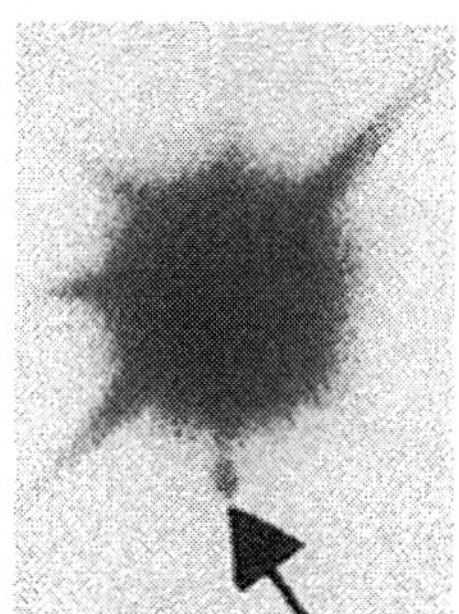

The arrow points to Sirius B, a white dwarf star.

Analysis of the orbit of the Sirius star system showed that the mass of the Pup was almost the same as that of our own Sun. This implied that Sirius B, a white dwarf, was thousands of times more dense than lead. Later, astronomers found that white dwarfs are common in our galaxy.

Brown Dwarf Stars

Brown dwarfs are substellar objects (~5 to 90 Jupiter masses) that do not fuse hydrogen into helium in their cores, as do stars on the main sequence, but have fully convective surfaces and interiors, with no chemical differentiation by depth. There is some question as to whether brown dwarfs are required to have experienced fusion at some point in their history; in any event, brown dwarfs heavier than 13 Jupiter masses do fuse deuterium.

Early stellar models suggested that a true star requires a mass at least 80 times that of Jupiter to support such fusion. Dense star-like objects with smaller masses, or "brown dwarfs," were hypothesized by the early 1960s -- formed much the way stars are formed, they would however be hard to find in the sky, as they would emit almost no light. Their strongest emissions would be in the infrared (IR) spectrum, and ground-based IR detectors were too imprecise for a few decades after that to firmly identify any brown dwarfs.

More recently, it has been hypothesized that, depending on the compounds that make up a growing stellar object, the critical mass for star-like hydrogen fusion could be as large as 90 Jupiter masses; and on the other end of the spectrum, that substellar objects formed quickly from a collapsing nebula could produce brown dwarfs smaller than 13 Jupiter masses, which nevertheless experience no fusion at all.

Since 1995, when the first brown dwarf was confirmed, hundreds have been identified. They are now believed to be the most numerous type of body in the Milky Way. Brown dwarfs close to Earth include Epsilon Indi Ba and Bb, a pair of dwarfs around 12 light-years from Sun.

Mainline stars cool, but eventually reach a minimum luminosity which they can sustain through steady fusion. This varies from star to star, but is generally at least 0.01% the luminosity of our Sun. Brown dwarfs cool and darken steadily over their lifetimes: sufficiently old dwarfs will be too faint to be a star.

A remarkable property of brown dwarfs is that they are all roughly the same radius, more or less the radius of Jupiter. At the high end of their mass range (60-90 Jupiter masses), the volume of a brown dwarf is governed primarily by electron degeneracy pressure, as it is in white dwarfs; at the low end of the range (1-10 Jupiter masses), their volume is governed primarily by Coulomb pressure, as it is in planets. The net result is that the radii of brown dwarfs vary by only 10-15% over the range of possible masses. This can make distinguishing them from planets difficult.

In addition, many brown dwarfs undergo no fusion; those at the low end of the mass range (under 13 Jupiter masses) are never hot enough to fuse even deuterium, and even those at the high end of the mass range (over 60 Jupiter masses) cool quickly enough that they no longer undergo fusion after something on the order of 10 million years. However, there are other ways to distinguish dwarfs from planets: Density is a clear giveaway. Brown dwarfs are all about the same radius and volume; so anything that size with over

10 Jupiter masses is unlikely to be a planet. X-ray and infrared spectra are telltale signs. Some brown dwarfs emit X-rays; and all "warm" dwarfs continue to glow tellingly in the red and infrared spectra until they cool to planet like temperatures (under 1000K).

Typical atmospheres of known brown dwarfs range in temperature from 300 to over 3000 K, in comparison with stars, which cool to minimum temperatures of around 4000 K. Compared to stars, which warm themselves with steady internal fusion, brown dwarfs cool quickly over time; more massive dwarfs cool slower than less massive ones.

Black holes

A black hole is a region of spacetime from which nothing can escape, even light. It is impossible to see a black hole directly because no light can escape from them; they are black. But there are good reasons to think they exist.

When a large star has burnt all its fuel, it explodes into a supernova. The stuff that is left from the explosion collapses down to an extremely dense object known as a neutron star. We know that these objects exist because several have been found using radio telescopes. If the neutron star is too large, the gravitational forces overwhelm the pressure gradients and collapse cannot be halted. The neutron star continues to shrink until it finally becomes a black hole. This mass limit is only a couple of solar masses that is about twice the mass of our sun, and so we should expect at least a few neutron stars to have this mass. (Our sun is not particularly large; in fact it is quite small.)

A supernova occurs in our galaxy once every 300 years, and in neighboring galaxies about 500 neutron stars have been identified. Therefore, we are quite confident that there should also be some black holes.

The Hubble Space Telescopic image given below is the center of galaxy NGC4261. Inside it, there is a brown spiral-shaped disk that is rotating. Based on gravitational calculations, it weighs hundred thousand times as much as our Sun. Physicists believe that there is a black hole at its center having gravity about one million times that of our Sun.

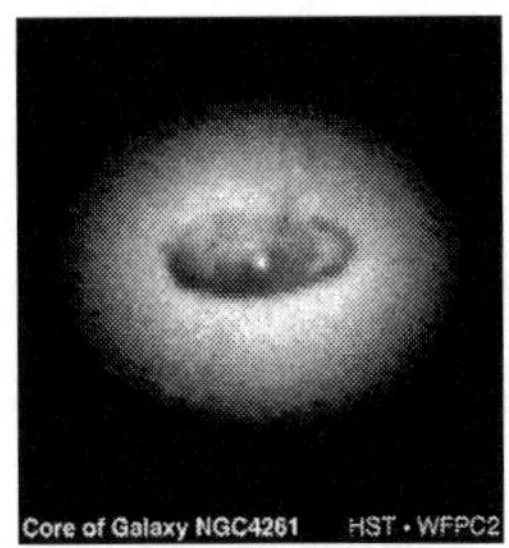

Hubble Space Telescope image of the center of galaxy NGC4261. Physicists believe that there is a black hole at its center.

Dark matters

Cosmologists believe that most of the matter in the universe is *dark*, i.e. cannot be detected from the light which it emits (or fails to emit). This is "stuff" which cannot be seen directly -- so what makes us think that it exists at all? Its presence is inferred indirectly from the motions of astronomical objects, specifically stellar, galactic, and galaxy cluster/supercluster observations. It is also required in order to enable gravity to amplify the small fluctuations in the Cosmic Microwave Background enough to form the large-scale structures that we see in the universe today.

For each of the stellar, galactic, and galaxy cluster/supercluster observations the basic principle is that if we measure velocities in some region, then there has to be enough mass there for gravity to stop all the objects flying apart. When such velocity measurements are done on large scales, it turns out that the amount of inferred mass is much more than can be explained by the luminous stuff. Hence we infer that there is dark matter in the Universe.

Dark matter has important consequences for the evolution of the Universe and the structure within it. According to general relativity, the universe must conform to one of three possible types: open, flat, or closed. The total amount of mass and energy in the universe determines which of the three possibilities applies to the universe. In the case of an open universe, the total mass and energy density (denoted by the Greek letter Omega) is less than unity. If the universe is closed, Omega is greater than unity. For the case where Omega is exactly equal to one the Universe is "flat".

Note that the dynamics of the Universe are not determined entirely by the geometry (open, closed or flat) unless the Universe contains only matter. In our Universe, where most of Omega comes from dark energy, this relation between the mass density, spatial curvature and the future of the universe no longer holds. It is then no longer true in this case that "geometry (spatial curvature) is destiny." Instead, to find out what will happen, one needs to calculate the evolution of the expansion factor of the universe for the specific case of matter density, spatial curvature and "funny energy" to find out what will happen.

Dark matter (DM) candidates are usually split into two broad categories, with the second category being further sub-divided into Baryonic, Non-Baryonic, hot dark matter (HDM), and cold dark matter (CDM) depending on their respective masses and speeds. CDM candidates travel at slow speeds (hence "cold") or have little pressure, while HDM candidates move rapidly (hence "hot").

Current indications from the cosmic microwave background are that the universe is spatially flat. That implies that the sum of all of the energy (density) in the universe equals the critical density, i.e. the total Omega is 1. This is quite interesting because as the universe expands the value of Omega changes. In fact the value 1 is unstable, and the universe would prefer to evolve towards one of the two natural values: 0, if the universe expands forever further apart until the universe is almost totally empty; and infinity, if the matter re-collapses to a state of higher and higher density. Then the observation that Omega is fairly close to 1 today, means that it must have been even closer to 1 in the past. It is unsatisfying to believe that we just happen to live at the time when Omega is just starting to depart from 1 by a small factor. It is much more appealing to consider that we do not live at a special epoch, so that Omega is still close to 1 today. But then we need to explain why Omega started out very close to 1 in the early universe. The theory of inflation provides just such a justification – most versions of inflation predict that the early universe was driven extremely close to flat, and that it is still very close to flat today. If this is so, then at least 90% the energy of the universe is dark! Note that although the universe may be flat that does not mean that matter makes up the critical density. In addition to dark matter there is dark energy, e.g. a cosmological constant that needs to be included in the accounting.

The cosmology of the expanded PM

The new cosmology offers a mechanical, not mathematical, interpretation of all universal phenomena and the universe consistent with PM. It is still the old 3-dimensional universe, where both time and dimensions are considered absolute but not light speed. As long predicted by scientists there is an atmosphere of aether (NEU) taking up all space of the universe to produce universal phenomena. Contrary to general belief and the teaching of no-aether theories, like a fire both the mass and the energy of the universe are not conserved. The stars of a universe constantly produce large amount of NEU, which continuously escape out of the universe carrying mass and energy. Besides, some of the outermost stars and galaxies are also escaping from the universe.

The host of the universe – Domain of the nature (DON)

A universe can be defined as a place where large amounts of stars live. Stars constantly undergo nuclear reactions releasing huge amount of electrons, protons, and energetic light-NEU. The entire universe is therefore an energy-consuming system having many engines (stars) running constantly burning fuel. There is always limited amount of fuel, thus, limited duration and size for any energy consuming system including a universe.

For a fire to burn there must be a bigger natural environment to make and store its fuel. For example, Earth provides an environment for plants to receive light energy from the Sun to be born, to grow old, and to die. A lightening triggers the starting of a tiny forest fire, which spreads and, eventually, when the stored forest fuel including the dead and the living plants, is exhausted, the fire dies. The energy-producing materials of a universe must first be produced and stored by some nature ways before it can be consumed. Also, there should be a nature mechanism to trigger the starting of the consumption of the stored energy-producing material. From what we know, the nature has designed a perfect engine inside all stars to continuously work for billions of years maintenance-free to burn the naturally stored energy-producing material by nuclear reactions at essentially constant rates!

A universe therefore needs to be in an even much bigger and longer-lasting natural environment to produce and to store energy-producing materials and to be born (to start consume the stored energy-producing material), to grow, and to die in it. The author calls this bigger environment the domain of the nature (DON). To understand our universe we need to figure out some basic properties about the DON; about what the energy-consuming material it produces and stores and what mechanism triggers the starting of consuming them or the birth of a universe.

The knowledge about the universe should be a good start point to figure out what the DON should be, which appears to have unlimited space and time, for example, but not the universe. There are likely many universes inside the DON obeying the same laws of physics and chemistry, existing simultaneously and/or at different time periods. A universe is separated from others by much bigger starless space than an intergalactic space (IS). The atmospheric NEU (neutrinos) of all universes continuously escape into the spaces separating the universes. Upon leaving a universe into the vast starless space of the DON, the atmospheric NEU continue to lose their kinetic energy and to reduce their concentrations to very low levels. There is likely a dilute low-energy atmosphere of NEU surrounding a universe and therefore there are the forces of DON, which are similar to the universal forces but much weaker. Both electrons and protons outside universes may still be charged particles having their unique structures.

In the DON outside universes the atmosphere of NEU should be unevenly distributed and there may be places in DON far away from any universe having close to real vacuum without NEU, electron and proton particles, even mini-aether in gas-like form.

Dense-matter object (DMO)

Both the concentration and the energy of the atmospheric NEU continue to decrease upon leaving a universe. The temperature in an IS is already very low. For example, the temperature in the IS surrounding our galaxy (Milky Way) is ~2.7 K. Upon leaving a universe, the temperature likely continues to drop to 0 K and below. Scientists have found that upon cooling to very close to 0 K, orbital electrons of atoms collapse into their nuclei to form a very dense matter yet unknown to us. The author has taken the liberty to name it the "dense-matter object (DMO)", which should contain about equal number of protons and electrons without atomic structure. The findings proved that outside universe in the DON where temperature is 0 K and below, atoms and molecules can no longer exist and will collapse to form DMO. Still under a dilute low-energy atmosphere of NEU or weak forces of the DON, DMOs would tend to agglomerate and continue to grow in size. Giving time and there is essentially unlimited supply of time in the DON outside universes, some DMOs can grow to very large sizes having amazingly large masses and relative kinetic energies among them.

Under very weak dilute atmosphere of NEU, protons and electrons may still be weakly charged and still have their own atmospheres of NEU. Low energy NEU may be even easier to be caught by the atmospheres of electrons and protons and they may be heavier and having denser atmospheres of NEU than those inside a universe. DMOs therefore may be made of protons and electrons having enriched NEU content, thus, may have even denser mass than atomic nuclei. DMOs therefore are formed and are stable in the DON outside universes under 0 K.

Having extremely dense mass, even under very dilute low-energy atmosphere of NEU, a large size DMO may still have quite strong gravitational force since they continue to collect NEU, mini-aether, protons, electrons, small size DMOs to grow in sizes. Making large sizes DMOs appears to be the nature way to concentrate both matter and their energy needed to make new galaxies and universes.

Having similar structure to atomic nuclei without orbital electrons to protect them, we would expect that DMOs are unstable inside a universe like radioactive nuclei. However, since they have amazingly dense mass, even the energetic atmospheric NEU inside a universe cannot penetrate deep into them. Therefore, only their surfaces are constantly been bombarded by energetic atmospheric NEU causing constant releasing electrons, protons, NEU along with huge amount of energy, thus, starting and maintaining constant-rate nuclear reactions only on the surfaces of DMOs. The DON therefore has the best nature way to concentrate mass and energy outside universes and the best sustained-release mechanism to utilize them inside universes!

Since MAM can not exist outside universes, a universe can not be born with them. Therefore, a universe should be born to contain 100% DMOs! All universes therefore must have an energetic atmosphere of NEU to make DMOs unstable undergoing nuclear reactions to make MAM and to release the huge amount of mass and energy stored in them. DMOs therefore are the starting materials for making galaxies and universes like ours, not hydrogen and helium.

Formation of a galaxy and a universe

Findings suggest that most stars in a galaxy appear to have about the same age indicating that they may be born simultaneously. It means that at the beginning, there were already billions of pieces of DMOs to begin nuclear reactions simultaneously on their surfaces to make hydrogen and helium. All universal phenomena are produced by collision interactions. The nature way to produce a galaxy therefore is likely a violent collision between two huge DMOs as illustrated in the picture given below.

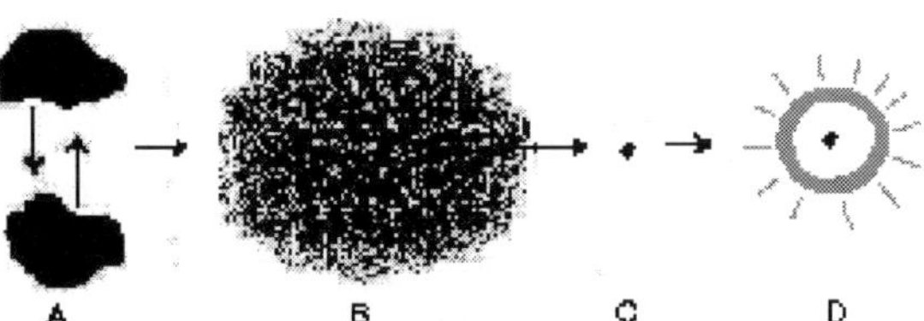

A violent collision of two huge DMOs (A) would break them into billions of pieces and produce a dense cloud containing dust of DMOs, electrons, protons, NEU, and huge amount of energy (B). Under constant bombardment of the energetic cloud particles, the broken pieces of DMOs and their dust particles quickly become "radioactive" starting nuclear reactions on their surfaces. The nuclear reactions have quickly consumed the dust and the small size DMOs producing hydrogen and helium clouds and an energetic atmosphere of NEU. Meanwhile, large pieces of DMOs quickly become baby stars by having continuous nuclear reactions on their surfaces to produce protons, electrons, hydrogen, helium, and NEU. The simultaneous formation of billions of baby stars marks the birth of a baby galaxy (B). The newly born baby stars are however essentially invisible since they are 100% DMOs, which without having atomic and molecular structures do not produce light except X-ray (B or C). An essentially invisible baby star containing 100% DMO is an antigravity star since it constantly shoots out energetic particles. Therefore, at the beginning a baby galaxy uniformly expands to spread their baby stars by their antigravity force pushing one another away. A baby star constantly throws out the hydrogen and the helium gas particles it continuously makes from its nuclear reactions to a great distance. Giving time the hydrogen and the helium particles gradually accumulate around a baby star at a great distance from its DMO with the help of their own gravitational and other physical and chemical forces forming an outer-layer. As the outer-layer grows thicker, an invisible baby star turns into a large bright young regular star (D).

Having gravitational force affecting one another, many large DMOs may exist in group in DON having orbiting-like motions. When two groups of DMOs run into each other, chances are many collisions will happen over a long period of time with each collision forming a galaxy. A universe may form this way over a rather long period of time. Galaxies in a universe therefore have different ages. Some are young and some old. Some may have been dead before a new galaxy is born. A universe therefore does not have a meaningful age but its galaxies do. A galaxy dies when its DMOs are about to be used up. One or a group of DMOs may run into a universe. A collision of a DMO with a star would produce unbelievably huge amount of energy and a fantastic view of explosion.

Stars

Stars cannot be formed by MAM such as hydrogen and helium since they cannot exist in the DON outside universes. Baby stars were born to be 100% DMOs instead. Under constant bombardment of energetic atmospheric NEU, DMOs inside a universe become radioactive constantly undergoing nuclear reactions to make hydrogen and helium. All stars undergo constant rate of nuclear reactions suggesting that these reactions only occur at their surface proving that even atmospheric NEU can not penetrate deep into DMOs due to their amazingly dense mass. Therefore, the mystery how stars can have nuclear reactions at their center at essentially constant rates possibly for billions of years without the need of concentrating hydrogen and removal of nuclear waste is now explainable.

Stars constantly produce hydrogen, helium, and heavier elements, by consuming their DMOs. The size and the shape of a DMO likely affect both the constant rate of nuclear reactions and the size of a star. A star dies when its DMO is nearly used up.

The protons and the electrons freed from DMOs by the bombardment of energetic NEU undergo nuclear reactions to form hydrogen and helium involving releasing X-Ray. DMOs or baby stars therefore should constantly radiate X-Ray and be visible by X-Ray telescope. However, most of them are too small and too far away to be detectable. Many DMOs may have irregular shapes and, therefore, producing imbalanced pushing force in all directions by their own radioactivity and nuclear reactions. As a result, many DMOs should have very strong rotational energy. Rotational DMOs creates spiral wind storms of NEU high in proton, electron, hydrogen, and helium contents around themselves, thus, having very strong magnetic force fields.

It is a universe made of both anti-gravitational and gravitational matters and forces not just gravitational force along. All DMOs are antigravity matter while all MAM such as planets are gravity matter. In general, a star continues to decrease its DMO content while to increase its MAM content upon aging, thus, turning from an antigravity star to a gravity star.

Regular stars

1. Having artificially low mass density

A baby star was born to 100% DMO, a strong antigravity star essentially invisible. It constantly makes hydrogen and helium throwing them out to a great distance to gradually accumulate forming its outer-layer. A regular young star therefore gradually forms a shinny outer layer containing hydrogen and helium constantly increasing in mass and brightness with aging. Constantly heated by the nuclear energy of its DMO this outer-layer is hot and bright. Since it is made of MAM, it should have gravity by reducing the energy of atmospheric NEU entering passing through it. A regular star is therefore made of an antigravity DMO having a outer-layer made of MAM having gravity. As a regular star gets older, the mass of its outer-layer continues to increase at the expense its DMO. It means that upon aging the gravity of a regular star's outer-layer gets stronger while the antigravity of its DMO gets weaker. Therefore, as a regular star ages it gradually turns from an antigravity star to a gravity star. A mature regular star like our Sun has already turned into an overall gravity star. Since the outer-layer of a regular star is at a great distance from its center where DMO is, it appears artificially very large and has artificially low mass density.

2. Developing explosive property

Since a DMO constantly throws out the hydrogen and the helium it makes, it should be the role of the outer-layer of a regular star to make elements heavier than helium. As the outer-layer gets thicker, it gets hotter and its inner surface gets even hotter. Besides, upon constant bombardment by energetic electrons, protons, hydrogen, helium, and X-ray-carrying NEU, the temperature on its inner surface will reach the critical point to begin nuclear reactions to make heavier elements than helium, sometime when a regular star reaches maturing age.

The outer-layer of a young regular star is readily penetrable and/or soluble to protons, electrons, hydrogen, and helium. However, as its inner-layer's content of heavier elements gets higher, it gets harder and harder for these particles to penetrate through and dissolve in. Therefore, as a regular star gets older, its outer-layer contains more elements heavier than helium particularly in the portion of the inner surface. As a result its inner gas pressure builds up. It will get to a point that a mature regular star can only periodically release its gas with bursts like the release of solar wind by our Sun. Also, with aging the DMO inside a star continues to get smaller, thus, allowing its outer layer to gradually cool off, thus, increasing its hardening. However, the DMO of a regular star continues to produce gas particles (hydrogen, helium, proton, electron), which continue to build up its inner pressure. A regular star therefore eventually will explode or violently expand becoming a red giant to release its inner pressure.

Dense stars

1. Black hole-like DMOs

Very large DMO may be present particularly at and near the center locations of galaxies. Their antigravity may be so strong that they throw their hydrogen and helium gases so far away that they constantly disperse instead of forming giant regular stars. Since their antigravity particles destroy and distort light, there is a large dark space surrounding them. This explanation of course counters the current concept of black holes having infinitely strong gravity, from which light cannot escape. Also some large DMOs have both very strong antigravity property and rotational energy and they may have very strong spiral wind storms of NEU around them to blow all their hydrogen and helium out to such a great distance to form giant bright rings having a donut-shape structure instead of forming a huge regular star leaving the DMO at its center invisible surrounded by dark space such as the image obtained by Hubble Space Telescope in the center of galaxy NGC4261NASA. Constantly pushed by the spiral neutrino wind storms produced by their DMOs, such ring structures rotate.

2. Neutron stars

The findings of neutron, white and brown dwarf stars, and black holes offer strong scientific evidences to prove the presence of DMOs in the universe. They are inside regular stars, exposed or partially exposed from the explosions of regular and white dwarf stars blowing all or part of their outer-layer away.

The new cosmology supports the teaching that large regular stars undergo supernovae explosion to form "neutron stars", which however should be exposed DMOs. A supernovae explosion would effectively remove the outer layer of a large regular star to expose its DMO becoming an invisible reborn-baby-star known as a "neutron star", which is by no means a dead star. A neutron star is a DMO, thus, is an antigravity baby star. It constantly produces hydrogen and helium throwing them out to a great distance to gradually form an outer layer and gradually becomes a shining regular star again.

Many fast rotating neutron stars have been found known to be Pulsars.

3. White and brown dwarfs

Medium-size stars such as our Sun may end their life by violently expansion to become red giants and then explode to shed a large portion of their outer layer leaving behind a new but much smaller outer-layer, which according to astronomical findings contains mostly carbon and oxygen. Under the influence of its own gravity, this new outer-layer is formed to have a minimum size balanced with the already reduced antigravity force of its DMO due to its smaller size. Both the closeness of the new outer-layer to its DMO and the high content of elements heavier than helium make the new stars very small, thus, very dense and hot. They are known to be "white dwarfs". Findings show that white dwarf and even brown dwarf stars are very small stars having mass densities much larger than those of regular stars and MAM. Findings also show that white dwarf stars are much hotter than ordinary stars. All findings prove that all dwarf stars contain DMOs to be dense stars. Having high content of elements heavier than helium, the outer-layer of a white dwarf is already not easily penetrable by the gas and the ion particles constantly produced by its DMO. Being extremely hot, the inner surface of the outer-layer should undergo nuclear reactions to produce even heavier elements than oxygen. As a result some white dwarfs are bound to violently explode again. They are known to undergo type Ia supernova explosion, which is the most powerful of all the supernovae. Currently, scientists believe that white dwarfs are dead stars having no nuclear fuel (hydrogen) left. Actually, they are very active stars undergoing nuclear reactions on both the surface of their DMOs and the inner surface of their outer-layer.

The DMOs inside brown dwarfs are even smaller than those in white dwarfs. The DMOs of larger brown dwarfs may be able to undergo nuclear reactions to produce hydrogen and helium but those of small brown dwarfs may be too small to undergo nuclear reaction but are always radioactive constantly producing large amount of energy. The antigravity of the DMOs of brown dwarfs may be too weak to separate themselves from their outer layers with a hallow space like regular stars and white dwarf stars. Once their DMOs have been used up, they become planets having low mass density containing mainly hydrogen and helium like Jupiter and Saturn.

No dark matter

Assuming that the movements of black-hole rings, galaxies and even the entire universe are controlled by their gravitational force alone, the mathematical models of modern cosmology requires the existence of very strong gravitational forces and, thus, much more masses than can be seen in the universe to produce them. Modern cosmology therefore has to postulate that the universe mainly contains invisible dark matters. According to the new cosmology, the rotation of black-hole rings is due to the spiral wind storms produced by their fast rotating DMOs while the movements of stars in galaxies and the entire galaxies are mainly controlled by NEU winds not gravity. There is no need to assume the presence of either dark matters or dark energies to explain all universal phenomena.

Solar and planetary motions

Having the ability of penetrating through MAM essentially freely, the drag effect of the atmospheric NEU is small for slow moving objects but it should still slow down all moving objects anyway. The drag effect is likely the reason why all satellites and space stations only stay in orbit for a few years and eventually fall back to Earth and for spaceships to slow down in space. The drag force of the atmospheric NEU however can be amazingly strong for very fast moving matters, for example, particles moving close to light speed need to encounter the atmospheric NEU moving at twice the light speed. It is therefore not surprising that scientists cannot accelerate electron or any other charge particles to light speed. They however believe the reason given by the theory of special relativity that mass increases with speed instead of simply the drag effect of the aether (atmospheric NEU).

One of the biggest secrets of the nature is that planets in the solar system have steady-state orbits around the Sun without either slowing down or falling into the Sun. It gives physicists the strongest reason to believe that there is no aether in space. Of course, it is also the most often raised question against the having-aether concepts. Actually, it is one of the most difficult scientific questions for scientists to answer with or without aether. Although the steady-state-like planetary motions are used to support no-aether theories, the solar system is just too orderly for scientists to conclude that it occurs naturally or randomly unless there is god or intelligent being to design it. Today's physicists support no-aether theories but they cannot explain such orderliness as that all nine planets orbit the Sun in the same direction and essentially on the same plain. In a broader scope, they cannot explain the movements of the Milky Way Galaxy and many other spiral galaxies.

In the presence of an atmospheric NEU, to have steady-state orbital motions planets need constant supply of energy to overcome both the drag effect of the atmospheric NEU and Sun's gravitational force. The only scientific possibility for them to do so is that the Sun somehow constantly provides the energy needed by all 9 planets and their moons to stay in their orbits. The expanded PM now offers such a scientific explanation given below.

There is a DMO inside all regular stars producing their nuclear reactions. The so-called neutron stars are actually exposed DMOs from supernova explosions of regular stars. DMOs are antigravity matters constantly undergoing radioactive-like activity and nuclear reactions shooting out energetic particles. Many neutron stars have been found fast rotating. Like many neutron stars the DMOs inside many stars such as our Sun are also fast rotating having very strong magnetic force fields around them. Since the DMO inside a regular star has a large space separating it from its outer layer, the fast rotation of its DMO does not affect the shape of its outer-layer, which we see. The outer-layer of a regular mature star like our Sun has strong gravity, which has overcome the antigravity of its DMO to become a gravity star.

The antigravity effects of the DMOs in many regular stars such as our Sun however have not been completely overcome by the gravity of its outer-layer in all directions due to their fast rotation. A fast rotating DMO creates a spiral NEU wind storm of NEU around it containing the gas particles it constantly produces including protons, electrons, hydrogen and helium. This storm should push its outer-layer to rotate. The outer-layer blocks most storm particles but not the spiral wind of NEU, which penetrates through the outer-layer to provide an antigravity force in the surrounding of the star. Although stars only produce light-energy-carrying NEU, there is always a dense atmosphere of undetectable NEU to constantly bombard the DMOs of stars. The spiral NEU wind therefore is made mainly by undetectable NEU, which should gain kinetic energy from the fast rotating DMO undergoing nuclear reactions. Only the spiral wind of undetectable NEU can easily penetrate the thick and hot outer-layer of stars to offer a directional anti-gravitational force to support all of their planetary motions.

If a planet is caught by the Sun in the direction its spiral NEU wind blow, it constantly gives the planet a push to maintain its orbiting motions. Such a spiral wind of NEU should have much stronger pushing force near the surface of the Sun and gets weaker quickly outward. Planets closer to the Sun, therefore, should orbit faster and need stronger push to overcome both the gravitational force of the Sun and the drag effect of the atmospheric NEU to stay in orbit than the planets farther away. Of course, all planets must orbit in the same direction of the spiral neutrino wind to survive. Contrary to general belief that some planets were formed during the formation of stars, all nine planets were caught by the Sun one at a time after it has become a gravity star. All planets should not orbit on the same plain to begin with. It must have taken a long period of time for all planets to gradually adjust their orbiting direction in the same direction the DMO of the Sun rotates resulting in a final equilibrium having all planets orbiting in the same plane.

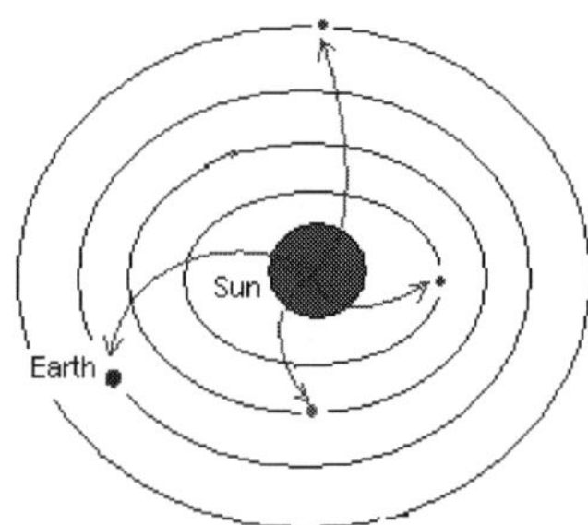

Our Sun produce a permanent spiral wind of NEU to push its outer layer to rotate and to keep its planets in steady-state orbits.

Since all the above given difficult nature moving conditions are met by all nine planets and the Sun, this scientific explanation has very strong scientific evidences to support it and becomes plausible. Once the Sun's gravity catches a new planet, it will take a long time for all its planets to reach a new equilibrium orbits. Finding no planet orbiting Sun in clockwise direction suggests that those planets have not survived. Statistically, there may be around equal number or nine planets and even more moons captured by Sun circulating in the wrong direction (clockwise) having plunged into Sun!

It is also reasonable to conclude that the same spiral neutrino wind keeps the moons of planets in orbits too. Most moons also orbit in the same counterclockwise direction around their planets with a few exceptions, which are the outer moons of Jupiter: Anake, Carme, Pasiphae and Sinope; of Saturn: Phoebe; of Neptune: Triton. However, they are all circulating the Sun counterclockwise.

As shown in the following picture the force of the spiral NEU wind of the Sun should have two components, the x-force in the planet orbiting direction to counter the drag of the atmospheric NEU and the anti-gravity force pushing planets away from Sun or the z-force. However, since Earth is in a steady-state orbit with all acting forces at equilibrium, we do no feel any force acting on it.

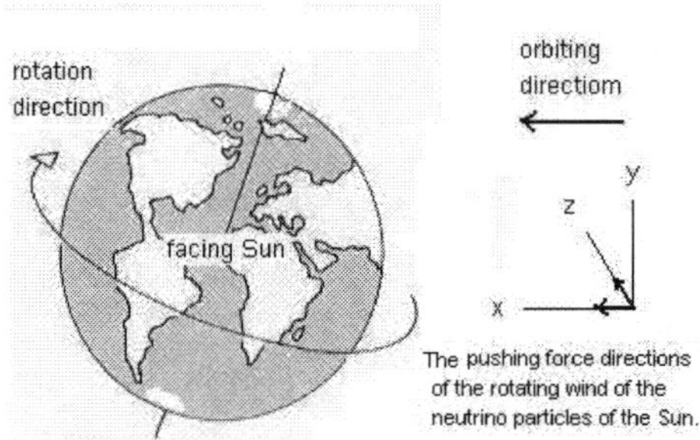

The spiral wind of NEU of the Sun has a pushing force on Earth in both directions, away from Sun (z-force) to counter gravity and in its orbiting direction (x-force) to keep Earth from slowing down.

A satellite or space station orbiting in the direction the Earth and the Moon orbiting around Sun should have longer lifetime than those orbiting in the opposite directions. In theory, those having maximized its utilization of the spiral NEU wind of the Sun should be able to stay in orbit essentially forever like what planets and their moons do.

The same spiral NEU wind of the Sun should also provide both the Sun and the planets with rotational energy. The storm particles, such as protons, electron, hydrogen and helium, of the spiral wind of NEU may provide even stronger force than the wind itself to push the outer-layer of the Sun to rotate. One possible mechanism to push planets to rotate is that the spiral wind of NEU has stronger pushing force on the half of the planets closer to the Sun than on the half farther away. If this is the correct mechanism, the larger a planet is, the larger the difference in the pushing forces on both halves, thus, the faster the planet should rotate. It would explain why the giant planets Jupiter and Saturn are the fastest rotating planets in the solar system!

There are however rare exceptions of the rules found. For example, Uranus and its moons rotate about its axis in a plane tilted 98 degrees with respect to the ecliptic and Venus rotates retrogradely and very slowly. These rare exceptions call for researches into special scientific reasons.

Winds of atmospheric NEU

Since all stars continuously produce NEU, each galaxy should produce a galaxy wind constantly blowing out in all directions. There are lots of galaxies so are their winds. When their winds encounter one another, many spiral or rotational winds of NEU, similar to hurricanes on Earth, may be routinely present in the universe.

All stars have a DMO at their center, which is made of protons and electrons similar to the nuclei of atoms but is much bigger. DMOs have very dense mass and even NEU cannot penetrate through them. As a result, the force of NEU wind can effectively act on DMOs, thus, stars. NEU winds therefore play a major role in affecting the movements of stars and galaxies particularly on very young galaxies and their young stars, which have very high DMO content. This is evident from the findings that there are many spiral galaxies, which are rotating. Astronomers have found that the movement of stars in a spiral galaxy does not fallow the rules of the movement of the planets in the solar system, which is believed to be a gravitational force-maintained system. For example, the stars closer to the center of the galaxy move at about the same velocity as those farther away from the center. These findings are strong scientific evidences that the movement of stars in a galaxy is mainly affected by the wind of NEU, not by gravitational force as believed by today's physicists.

There are many spiral-shape galaxies explainable by that their movements are mainly controlled by NEU winds.

Hurricanes are formed with spiral winds and their shapes are amazingly similar to those of spiral galaxies. Besides, they are also similar in their behaviors too. For example, the wind speed inside a hurricane is about the same and therefore its speed can specify its strength.

A hurricane is shaped by spiral wind so are many galaxies.

Since the star distribution in the universe is uneven, it is possible that the average concentration or even energy of the atmospheric NEU around Earth or the solar system may change over time. The change may affect our weather pattern and even volcanic activities.

Chapter 5 – Overview and discussion

This chapter broadly discusses the expanded PM in relation to other theories and to important scientific issues. It is time to confront and solve very controversial scientific issues lasted for several centuries already.

The main issue

There have long been many physics contradictory to one another. Among them only PM has been repeated proven correct over several centuries and teaches that all phenomena must have physical origins meaning that they are produced by matters having masses and relative motions. All other theories are mathematical theories seeking mathematical origins to interpret and to represent all universal phenomena.

Both modern physics and modern cosmology are based on Einstein's theory of relativity, which is known as the mainstream of thought and a no-aether theory. Einstein's theory of relativity was further based on Maxwell Equations derived from the postulations that force fields existed. Many physicists disagreed with Einstein's no-aether interpretation of Maxwell Equations leading to plenum aether theory. The plenum aether theory is therefore a twin of Einstein's theory of relativity born from Maxwell Equations sharing the same postulations of the existence of force fields and their mathematical consequences such as the invariance of light speeds, variance of time and dimensions, and their paradoxes. Plenum aether theory is contradictory with both PM and the mainstream of thought. Mainstream scientists so far have ignored plenum-aether theory and suggested that PM was only validity within a range. According to them PM is neither valid in environments having relativistic-velocities matters nor in microenvironments, which are quantized.

Only one of all the physics contradictory with one another can be correct, which should also be the physics of everything. Whichever can offer an overall interpretation of the universe and all its universal phenomena supported by all findings should be the only correct physics. Now, competition is over since the expanded PM has been proven to be the only correct physics or the physics of everything.

NEU v. neutrinos

Current theory of neutrino and findings

Neutrinos are elementary particles that travel close to the speed of light. Lacking an electric charge, they are able to pass through ordinary matter almost undisturbed and are thus extremely difficult to detect. As of 1999, it is believed neutrinos have a minuscule, but nonzero mass. They are created as a result of certain types of radioactive decay or nuclear reactions such as those that take place in the Sun, in nuclear reactions, or when cosmic rays hit atoms. There are three types, or "flavors", of neutrino: electron neutrino, muon neutrino and tau neutrino. Electron neutrinos or antimatter partner, called an antineutrino. Electron neutrinos or antineutrinos are generated whenever neutrinos change into protons or vice versa and by the two forms of beta decay. Interactions involving neutrinos are generally mediated by the weak forces.

Most neutrinos passing through the Earth emanated from the Sun, and more than 50 trillion solar electron neutrinos pass through the human body every second. The neutrino was first postulated in December 1930 by Wolfgang Pauli to preserve conservation of energy, conservation of momentum, and conservation of angular momentum in beta decay, the decay of a neutron into a proton, an electron and an antineutrino. Pauli theorized that an undetected particle was carrying away the observed difference between the energy, momentum, and angular momentum of the initial and final particles.

In 1942 Kan-Chang Wang first proposed to use beta-capture to experimentally detect neutrinos. In 1956 Clyde Cowan, Frederick Reines, F. B. Harrison, H. W. Kruse, and A. D. McGuire published the article "Detection of the Free Neutrino: a Confirmation" in Science, a result that was rewarded with the 1995 Nobel Prize. In this experiment, now known as the neutrino experiment, neutrino created in a nuclear reactor by beta decay were shot into protons producing neutrons and positrons both of which could be detected. It is now known that both the proposed and the observed particles were antineutrinos.

In 1962 Leon M. Lederman, Melvin Schwartz and Jack Steinberger showed that more than one type of neutrino exists by first detecting interactions of the muon neutrino (already hypothesised with the name of neutretto), which earned them the 1988 Nobel Prize. When a third type of lepton, the tau, was discovered in 1975 at the Stanford Linear Accelerator, it too was expected to have an associated neutrino. First evidence for this third neutrino type came from the observation of missing energy and momentum in tau decays analogous to the beta decay leading to the discovery of the neutrino. The first detection of tau neutrino interactions was announced in summer of 2000 by the DONUT collaboration at Fermilab, making it the latest particle of the Standard Model to have been directly observed; its existence had already been inferred by both theoretical consistency and experimental data from LEP.

Starting in the late 1960s, several experiments found that the number of electron neutrino arriving from the Sun was between one third and one half the number predicted by the Standard Solar Model, a descrepency which became known as the solar neutrino problem and remained unsolved for some thirty years, The Standard Model of particle physics assumes massless neutrino that don't change flavor. However, nonzero neutrino mass and accompanying flavor oscillation remained a possibility.

A practical method for investigating neutrino masses (that is, flavor oscillation) was first suggested by Bruno Pontecorvo in 1957 using an analogy with the neutral kaon system; over the subsequent 10 years he developed the mathematical formalism and the modern formulation of vacuum oscillation. In 1985 Stanislav Mikheyev and Alexei Smirnov (expanding on 1978 work by Lincoln Wolfenstein) noted that flavor oscillations can be modified when neutrinos propagate through matter. This so-called MSW effect is important to understand neutrinos emitted by the Sun, which pass through its dense atmosphere on their way to detectors on Earth.

Starting in 1998, experiments began to show that solar and atmospheric neutrinos change flavors (see Super-Kamiokande, Sudbury Neutrino Observatory). This resolved the solar neutrino problem: the electron neutrino produced in the Sun had partly changed into other flavors which the experiments could not detect.

Although individual experiments, such as the set of solar neutrino experiments, are consistent with non-oscillatory mechanisms of neutrino flavor conversion, taken altogether, neutrino experiments imply the existence of neutrino oscillations. Especially relavent in this context are the reactor experiment KamLAND and the accelerator experiment such as MINOS. The KamLAND experiment has indeed identified oscillation as the neutrino flavor conversion mechanism involved in the solar electron neutrino. Similarly MINOS confirms the oscillation of atmospheric neutrinos and gives a better determination of the mass squared splitting (Maltoni, 2004).

Raymond Davis Jr. and Masatoshi Koshiba were jointly awarded the 2002 Nobel Prize in Physics. Ray Davis for his pioneer work on cosmic neutrinos and Koshiba for the first real time observation of Supernova neutrinos. The detection of Solar neutrinos, and of neutrinos of SN 1987A supernova in 1987 marked the beginning of neutrino astronomy.

An experiment done by C. S. Wu at Columbia University showed that neutrinos always have left-handed chirality. Detection of neutrino is challenging and often requires large detection volume or high intensity artificial neutrino beam. To date, not very much is experimentally established concerning the interactions of neutrinos with matter.

It is very hard to uniquely identify neutrino interactions among the natural background of radioactivity. For this reason, in early experiments a special reaction channel was chosen to facilitate the identification: the interaction of an antineutrino with a hydrogen neucleus, which is a single proton. It turns out that an antineutrino would travel about 30 light years through water before it undergoes this specific reaction. This doesn't prove in any way that an antineutrino cannot undergo other reactions with matter. But however it led to popular misconceptions like this one:

Because the cross section in weak nuclear interactions is very small, neutrino can pass through matter almost unhindered. For typical neutrino produced in the sun (with energies of a few MeV), it would take approximately one light year ($\sim 10^{16}$ m) of lead to block half of them. For example, in later theories, such as the one describing a so called MSW effect, it is thought that most solar neutrinos are interacting with matter inside the sun.

It is possible that the neutrino and antineutrino are in fact the same particle, a hypothesis first proposed by the Italian physicists Ettore Majorana. The neutrino could transform into an antineutrino (and vice vera) by flipping the orientation of its spin state.

The neutrinos from the supernova 1987a were detected within a time window that was consistent with a speed of light for the neutrinos.

Neutrino Mass – The Standard Model of particle physics assumed that neutrinos were massless, although adding massive neutrino to the basic framework is not difficult. Indeed, the experimentally established phenomenon of neutrino oscillation requires neutrino to have non-zero masses.

The strongest upper limit on the masses of neutrino comes from cosmology : the Big Bang model predicts that there is a fixed ratio between the number of neutrino and the number of photons in the cosmic microwave background. If the total energy of all three types of neutrinos exceeded an average of 50 electronvolts per neutrino, there would be so much mass in the universe that it would collapse. This mass can be circumvented by assuming that the neutrino is unstable; however, there are limits within the Standard Model that make this difficult. A much more stringent constraint comes from a careful analysis of cosmological data, such as the cosmic microwave background radiation, galaxy surveys and the Lyman-alpha forest. These indicate that the sum of the neutrino masses must be less than 0.3 electronvolt (Goobar, 2006).

In 1998, research result at the Super-Kamiokande neutrino detector determined that neutrinos do indeed flavor oscillate, and therefore have mass. The experiment is only sensitive to the difference in the squares of the masses (Mohapatra, 2005). The best estimate of the difference in the squares of the masses of mass eigenstate 1 and 2 was published by KamLAND in 2005: $\Delta m_{21}^2 = 0.000079$ eV^2. In 2006, the MINOS experiment measured oscillation from an intense muon neutrino beam, detrmining the difference in the squares of the masses between neutrino mass eigenstates 2 and 3. the initial results indicate $\Delta m_{23}^2 = 0.003$ eV^2, consistent with previous result from Super-K.

Experimental results show that (nearly) all produced and observed neutrinos have left-handed helicities (spins antiparallel to momenta), and all antineutrinos have right-handed helicities, within the margin of error. In the massless limit, it means that only one of two possible chiralities is observed for either particle. These are the only chiralities included in the Standard Model of particle interactions.

It is possible that their counterparts (right-handed neutrino and left-handed antineutrino) simply do not exist. If they do, their properties are substantially different from observable neutrino and antineutrino. It is theorized that they are either very heavy (on the order of GUT scale—see Seesaw Mechanism), do not participate in weak interaction (so-called sterile NEU), or both.

The existence of nonzero neutrino masses somewhat complicates the situation. Neutrinos are produced in weak interactions as chirality eigenstates. However, chirality of a massive particle is not a constant of motion; helicity is, but the chirality operator does not share eigenstates with the helicity operator. Free neutrino propagate as mixtures of left- and right-handed helicity states, with mixing amplitudes on the order of m_ν / E. This does not significantly affect the experiments, because neutrinos involved are nearly always ultrarelativistic, and thus mixing amplitudes are vanishingly small (for example, most solar neutrinos have energies on the order of 100 keV–1 MeV, so the fraction of neutrino with "wrong" helicity among them cannot exceed 10^{-10}).

Nuclear reactors are the major source of human-generated neutrinos. Anti-neutrinos are made in the beta-decay of neutron-rich daughter fragments in the fission process. Generally, the four main isotopes contributing to the anti-neutrino flux are: uranium-

235, uranium-238, plutonium-239, plutonium-241 (e.g. the anti-NEU emitted during beta-minus decay of their respective fission fragments). The average nuclear fission releases about 200 MeV of energy, of which roughly 6% (or 9 MeV, depending on quoted reference) are radiated away as anti-neutrino. For a typical nuclear reactor with a thermal power of 4,000 MW (Megawatts) and an electrical power generation of 1,300 MW, this corresponds to a total power production of 4,250 MW, of which 250 MW is radiated away, and disappears, as anti-neutrino radiation. This is to say, 250 MW of fission energy is *lost* from this reactor and does not appear as heat, since the anti-neutrino penetrate all normal building materials essentially tracelessly. The exact energy spectrum is mostly uncertain and depends, for example, on the degree to which the fuel is burned.

There is no established experimental method to measure the flux of low energy anti-neutrinos. Only anti-neutrinos with an energy above threshold of 1.8 MeV can be uniquely identified (see "Neutrino Detection" below). An estimated 3% of all anti-neutrinos from a nuclear reactor carry an energy above this threshold. An average nuclear power plant may generate over 10^{20} anti-neutrinos per second above this threshold, and a much larger number which cannot be seen with present detector technology.

Some particle accelerators have been used to make neutrino beams. The technique is to smash protons into a fixed target, producing charged pions or kaons. These unstable particles are then magnetically focused into a long tunnel where they decay while in flight. Because of the relativistic boost of the decaying particle the neutrinos are produced as a beam rather than isotropically. Nuclear bombs also produce very large quantities of neutrinos. Fred Reines and Clyde Cowan considered the detection of neutrinos from a bomb prior to their search for reactor neutrinos.

Neutrinos are also produced as a result of natural background radiation. In particular, the decay chains of uranium-238 and thorium-232 isotopes, as well as potasium-40, include beta decays which emit anti-neutrinos. These so-called geoneutrinos can provide valuable information on the Earth's interior. A first indication for geoneutrino was found by the KamLAND experiment in 2005. KamLAND's main background in the geoneutrino measurement are the anti-neutrino coming from reactors. Several future experiments aim at improving the geoneutrino measurement and these will necessarily have to be far away from reactors.

Atmospheric neutrino result from the interaction of cosmic rays with atomic nuclei in the Earth's atmosphere, creating showers of particles, many of which are unstable and produce neutrino when they decay. A collaboration of particle physicists from Tata Institute of Fundamental Research (TIFR), India, Osaka City University, Japan and Durham University, UK recorded the first cosmic ray neutrino interaction in an underground laboratory in KGF gold mines in India in 1965.

Solar neutrinos originate from the nuclear fusion powering the sun and other stars. The details of the operation of the sun are explained by the Standard Solar Model. In short: when four protons fuse to become one helium nucleus, two of them have to convert into neutrons, and each such conversion releases one electron neutrino.

The sun sends enormous numbers of neutrinos in all directions. Every second, about 70 billion (7×10^{10}) solar neutrino pass through every square centimeter on Earth that faces the sun. Since neutrinos are insignificantly absorbed by the mass of the Earth, the surface area on the side of the Earth opposite the Sun receives about the same number of neutrinos as the side facing the Sun.

Neutrinos are an important product of Types Ib, Ic and II (core-collapse) supernovae. In such events, the pressure at the core becomes so high (10^{14} g/cm^3) that the degeneracy of electrons is not enough to prevent protons and electrons from combining to form a neutron and an electron neutrino. A second and more important neutrino source is the thermal energy (100 billion kelvins) of the newly formed neutron core, which is dissipated via the formation of neutrino-antineutrino pairs of all flavors. Most of the energy produced in supernovas is thus radiated away in the form of an immense burst of neutrino. The first experimental evidence of this phenomenon came in the year 1987, when neutrino from supernova 1987A were detected. The water-based detectors Kamiokande II and IBM detected 11 and 8 antineutrino of thermal origin, respectively, while the Gallium-71-based Baksan detector found 5 neutrino (lepton number = 1) of either thermal or electron-capture origin, in a burst lasting less than 13 seconds. It is thought that neutrino would also be produced from other events such as the collision of neutron stars. What was particularly interesting about this event was that the neutrino signature of the supernova arrived at earth approximately 18 hours before the arrival of the first photon signature. The exceptionally weak interaction with normal matter allowed the neutrino to pass through the churning mass of the exploding star, while the electromagnetic photons were retarded, with the photon signature of the supernova not being released until the outermost layers of the star were superheated and released a much brighter visible light signature, observed telescopically on earth some 18 hours after the neutrino had already arrived. This point shows how weakly interacting neutrino truly are.

Because neutrinos interact so little with matter, it is thought that a supernova's neutrino emissions carry information about the innermost regions of the explosion. Much of the visible light comes from the decay of radioactive elements produced by the supernova shock wave, and even light from the explosion itself is scattered by dense and turbulent gases. Neutrinos, on the other hand, pass through these gases, providing information about the supernova core (where the densities were large enough to influence the neutrino signal). Furthermore, the neutrino burst is expected to reach Earth before any electromagnetic waves, including visible light, gamma rays or radio waves. The exact time delay is unknown, but for a Type II supernova, astronomers expect the neutrino flood to be released seconds after the stellar core collapse, while the first electromagnetic signal may be hours or days later. The SNEWS project uses a network of neutrino detectors to monitor the sky for candidate supernova events; it is hoped that the neutrino signal will provide a useful advance warning of an exploding star.

The energy of supernova neutrino ranges from a few to several tens of MeV. However, the sites where cosmic rays are accelerated are expected to produce neutrino that are one million times more energetic or more, produced from turbulent gasesous environments left over by supernova explosions: the supernova remnants. The connection between cosmic rays and supernova remnants was suggested by Walter Baade and Fritz Zwicky, shown to be consistent with the cosmic ray losses of the Milky Way if the efficiency of acceleration is about 10 percent by Ginzburg and Syrovatsky, and it is supported by a specific mechanism called "shock wave acceleration" based on Fermi ideas (which is still under development). The very high energy neutrino are still to be seen, but this branch of neutrino astronomy is just in its infancy. The main existing or forthcoming experiments that aim at observing very high energy neutrino from our galaxy are Baikal, AMANDA, [ICECUBE], Antares, NEMO and Nestor. Related information is provided by very high energy gamma ray observatories, such as HESS and MAGIC. Indeed, the collisions of cosmic rays are supposed to produce charged pions, whose decay give the neutrino, but also neutral pions, whose decay give gamma rays: the environment of a supernova remnant is transparent to both types of radiation.

Still higher energy neutrino, resulting from the interactions of extragalactic cosmic rays, could be observed with the cosmic ray observatory Auger or with the dedicated experiment named ANITA.

It is thought that, just like the cosmic microwave background radiation left over from the Big Bang, there is a background of low energy neutrino in our Universe. In the 1980s it was proposed that these may be the explanation for the dark matter thought to exist in the universe. Neutrinos have one important advantage over most other dark matter candidates: we know they exist. However, they also have serious problems.

From particle experiments, it is known that neutrinos are very light. This means that they move at speeds close to the speed of light except when they have extremely low kinetic energy. Thus, dark matter made from neutrino is termed "hot dark matter". The problem is that being fast moving, the neutrino would tend to have spread out evenly in the universe before cosmological expansion made them cold enough to congregate in clumps. This would cause the part of dark matter made of neutrino to be smeared out and unable to cause the large galactic structures that we see. Further, these same galaxies and groups of galaxies appear to be surrounded by dark matter which is not fast enough to escape from those galaxies. Presumably this matter provided the gravitational nucleus for formation. This implies that neutrino make up only a small part of the total amount of dark matter.

From cosmological arguments, relic background neutrino are estimated to have density of 56 of each type per cubic centimeter and temperature 1.9 K (1.7×10^{-4} eV) if they are massless, much colder if their mass exceeds 0.001 eV. Although their density is quite high, due to extremely low neutrino cross-sections at sub-eV energies, the relic neutrino background has not yet been observed in the laboratory.

Because neutrinos are very weakly interacting, neutrino detectors must be very large in order to detect a significant number of neutrinos. Neutrino detectors are often built underground in order to isolate the detector from cosmic rays and other background radiation. Antineutrinos were first detected in the 1950s near a nuclear reactor. Reines and Cowan used two targets containing a solution of cadmium chloride in water. Two scintillation detectors were placed next to the cadmium targets. Antineutrino with an energy above the threshold of 1.8 MeV caused charged current interactions with the protons in the water, producing positrons and neutrons. The resulting positron annihilations with electrons created photons with an energy of about 0.5 MeV. Pairs of photons in coincidence could be detected by the two scintillation detectors above and below the target. The neutrons were captured by cadmium nuclei resulting in gamma rays of about 8 MeV that were detected a few microseconds after the photons from a positron annihilation event.

Since then, various detection methods have been used. Super Kamiokande is a large volume of water surrounded by photomultiplier tubes that watch for the Cherenkov Radiation emitted when an incoming neutrino creates an electron or muon in the water. The Sudbury Neutrino Observatory is similar, but uses heavy water as the detecting medium, which uses the same effects, but also allows the additional reaction any-flavor neutrino photo-dissociation of deuterium, resulting in a free neutron which is then detected from gamma radiation after chlorine-capture. Other detectors have consisted of large volumes of chlorine or gallium which are periodically checked for excesses of argon or germanium, respectively, which are created by electron-neutrino interacting with the original substance. MINOS uses a solid plastic scientilator coupled to photomultiplier tubes, while Borexino uses a liquid pseudocumene scintillator also watched by photomultiplier tubes while the proposed NovA detector will use liquid scintillator watched by avalanche photodiodes.

NEU or neutrinos

Although it has been over half century since the postulation of the existence of neutrinos and the proof of its existence indirect, the neutrino science is still an unproven theory. Although scientists believe the presence of a dense atmosphere of neutrinos, they do not believe neutrinos have anything to do with universal phenomena. Earlier, they thought that neutrinos do not interact with MAM at all but now they are changing their mind and suggest that neutrinos do collide interact with the MAM of Sun.

For neutrinos (NEU) the scientific community has returned to the era when atoms and molecules are undetectable directly but their existence was expected from many physical and chemical evidences. The difference is that this time physicists have already given up the hope of a mechanical or physical interpretation of universal phenomena. Their wrong ideas mainly came from Their long-time misunderstanding of charge energy.

Physicists are very sure that all stars constantly produce a lot of neutrinos. Yet, they are still blinded by their own theories to the following facts. 1. neutrinos are the only material particles taking up all the space of the universe including the spaces occupied by all MAM constantly replenished by stars. 2. They have the unique physical properties to explain all universal phenomena. 3. They have the energy supply from stars to produce all universal phenomena. 4. They are constantly released from nuclear reactions proving that electrons and/or protons contain them. If the presence of neutrinos can explain only one universal phenomena, it may be a coincidence. However, if it explains all universal phenomena, it has to be correct!

NEU should be neutrinos produced by stars from their nuclear reactions. They are however very different theoretically since NEU has to be very interactive with MAM to produce all universal phenomena. Yet, scientists believe that neutrinos are very inactive only interacting rarely with MAM, since they have concluded from their observations that neutrinos pass through MAM without producing heat. Now, the expanded PM explains why NEU or neutrinos are very interactive but their collision interactions with MAM essentially do not produce heat.

The mainstream of thought (MOT)

Since the acceptance of Newton's theory of universal gravitation in 18th Century the entire scientific community has gradually turned to no-aether theories believing that all matters have gravitational force, which is an attraction force acting from distance.

Mainstream view is mainly based on Einstein's theories of special and general relativity also including big bang theory and quantum theory. Based on the assumption of invariance in light speed, Einstein mathematically concluded that time and dimensions are functions of environmental mass content and distribution and relative velocity of matters. Based on Hubble's findings of both universal redshift of star lights and Hubble Law MOT teaches that the universe began with a big bang of something extremely small, dense, and hot known as "singularity" about 15 billion years ago and has been uniformly expanding ever since.

Einstein's world is a four-dimensional universe filled with the so-called matter-free spacetime instead of aether. Although spacetime has no mass, it somehow can be distorted or curved by matters (their one-sided interaction with spacetime) and their moving speed. Also, somehow, curved spacetime produces gravity, gravitational redshift, and even bends lights. It teaches a gravitational force-shaped universe where gravity controls all stellar motions and crashes MAM to form dense stars and black holes. It teaches that the motions in the solar system are controlled by gravitational forces so are the motions of stars in galaxies. Calculations however have concluded that the total visible masses of rotational galaxies are too small to produce strong enough gravitational forces to prevent their stars from flying apart. Therefore, mainstream physicists and cosmologists have postulated the presence of invisible or dark matters, which, according to their calculation, should be much more than all visible matters.

According to MOT the universe is uniformly expanding. Since gravitational force is the only force to determine the fate of the universe, it may expand forever (open), eventually stop expanding (flat), or begin to contract after stopping expanding (close) depending on the mass content of the universe.

In the mainstream universe light speed is invariant but time, dimension, and mass are not. In the so-called relativistic environment high relativistic speeds increase mass and slow down time, where mainstream scientists believe that PM laws fail and should be replaced by theories of relativity. Finding that atomic and molecular spectra (energies) are quantized, Planck developed quantum theory suggesting that energies are quantized in microenvironments, which has also been included in MOT. In modern physics quantum theory is as important as theories of relativity particular in particle physics. Again, the mainstream physicists insist that in micro environment the laws of PM fail, which should be replaced by quantum theory. Since gravitational force is not quantized, theories of relativity are basically inconsistent with quantum theory.

Both difficult mathematics and scientific incomprehensibility have led the public to believe that only very few physicists (rocket scientists) can understand (modern) physics and cosmology. Can they really? To be correct MOT should be able to explain everything but Einstein and many others have tried to develop a MOT-based theory of everything without success. Presently, some physicists are developing and advertise a theory called string theory suggesting that the universe has 11 dimensions. In the mathematical world there are infinite dimensions but in the physical world only three dimensions make sense.

A century after Einstein's theories of relativity, both universal phenomena and the universe are still as mysterious as ever. For example, still nobody really knows what gravity, light, and other universal phenomena and even the universe are. If the mainstream views were correct, it should be able to interpret all universal phenomena logically, coherently, and consistently and the predicted aether or the material particles to produce them should never be found. Now, both the failure of the efforts of Einstein and many physicists to develop a theory of everything and the discovery of the expanded PM have disproved MOT including both modern physics and modern cosmology.

The energy concepts of MOT are based on the famous Einstein Equation. Under the assumptions of mass increasing with traveling speed, invariance of light speed, and that nothing can travel at or faster than light, Einstein derived his famous equation $E = mc^2$. It further led to the postulations that matter-free energy exists, that mass and energy were mutually convertible. The findings that a small portion of mass is lost in nuclear reactions have been used to prove Einstein Equation suggesting that the lost mass has been converted to energy producing huge amount of nuclear energy obeying Einstein Equation. Finding positron has further led to the postulations of the presence of antiparticles, antimatters and that the collision between a particle and its antiparticle results in annihilating them into pure energy again obeying Einstein Equation.

The misled concept from Newton's gravity theory that universal forces are attraction forces, which can act from distance has led to Faraday's force-field theory. This theory has led to both the no-aether theories or MOT and the plenum aether theories. Mathematical theories are all based on the postulation of the existence of matter-free or non-physical force fields. If we still do not know the presence of an air atmosphere to produce wind, atmospheric pressure, fire, etc., the postulation of the presence of force fields to mathematically interpret these phenomena certainly cannot led to scientific understanding of them and the discovery of air.

According to the expanded PM a force field must have a physical origin meaning that it has to be a particle field. The acceptance of postulations and theories violating the fundamental physics laws of PM such as the laws of mass-balance and conservation of energy has led the world into an era of religious pseudo-physics and cosmology for over a century already.

The solar system

The origin and the age questions

In the beginning of the universe no MAM exists. Scientists should be correct that planets like Earth and Mars are formed from the remnants of stars' explosions. However, planets containing mainly hydrogen and helium are likely formed from small DMOs not from the spinning-off during the star-formation by hydrogen and helium gases as scientists have believed. It means that all stars do not have planet to start with. Besides, they are antigravity stars to start with too. All the planets of our Sun are therefore captured one at a time after it has become a gravity star having only a directional antigravity.

Using radiometric-dating method scientists found that (cooled down) Earth is about 3.5 billion year old and the ages of the meteorites found on Earth are around 4.5 billion year old. Earth should be produced by the explosion of a regular star. It is unknown how long a life cycle a regular star (to form, to age, and to finally explode) to produce both planets and meteors. Taking this period of time into account, Earth and other planets, besides the hydrogen and helium containing planets, are likely even much older than estimation.

Our Sun is a typical mature regular star. It is likely making elements heavier than helium on the inner surface of its outer-layer making it harder and harder for the gases and ions constantly made by its DMO to dissolve in and penetrate through. Some day it will violently expand and/or explode to end its life together with the civilization on Earth. Since big bang theory is likely wrong nobody really knows when this will happen!

Global temperature changes

Earth is very unique to have an environment suitable for the existence of biological lives. Having the correct narrow-range of temperature, an atmosphere of air and plenty of water is the necessary environmental conditions for the existence of biological lives. No biological life was found in the remaining planets of the solar system. There may be billions of planets in the universe but it is still doubtful whether there is a second Earth having biological lives.

On Earth even a few degree change in average temperature would lead to environmental disasters, since biological lives and its environment cannot tolerate significant temperature changes. The solar system must have gone through a long period of time with all the planets of the solar system having close to steady-state orbits and having constant energy output by our Sun. Still the Earth has been in and out of ice ages in a change cycle of about 100,000 years in the past 400,000 years having its average surface temperature varying in a range of about 12 °C. According to this historic trend, we are again in the middle of the warm or interglacial period. Scientists have found that changes in the shapes of the Earth's orbit are likely the main reasons for the above periodical temperature fluctuations. Its eccentricity changing cycle is about 100,000 years, tilt cycle is 41,000 years and its precession cycle is between 19,000 to 23,000 years. Besides, scientists think both changing in Sun's intensity and Earth's volcanic eruptions are the other important factors affecting Earth's average surface temperature.

One of the hottest debates, both scientific and political, is about the global warming trend increasing about 1.5 °F in the past century. During this period, human activities have changed the composition of Earth's atmosphere by increasing the levels of greenhouse gases, for example, from ~ 280 ppm in pre-industrial time to 382 ppm in 2006 for carbon dioxide content in air. Rising greenhouse gases in the atmosphere many cause global warming. Many scientists suggest that as the content of greenhouse gases in air continues to increase in the coming decades both the average global temperature and sea level will continue to rise. The issue whether the temperature rise of last century has crossed over the warming limit of the boundary has become a hot and controversial topic. Many scientists and politicians are demanding governmental actions to drastically reduce fossil fuel consumption. Without clean energies ready to replace fossil fuels, the world will suffer hardship and economical recession from these actions.

Nevertheless, it is fortunate to have raised the awareness that global-temperature-change is such an important scientific issue that the future existence of mankind is at steak. Hopefully, it will lead to seeking a real scientific solution to save the future of mankind. We do have a manmade global warming problem and stop using fossil fuel may be the correct solution, which however is a very specific solution useless for solving other nature temperature change problems such as the ice-age problem we shall face in the future. Global weather change, warming or cooling, will bring disaster and threaten the very existence of human race. Besides solving fossil fuel caused global warming problem, we therefore need a non-specific scientific solution for both global warming and cooling problems arising from all natural causes.

We naively believe that all stars are gravity stars and the gravity of the Sun should remain constant essentially forever. We believe that the orbits of all planets should remain unchanged until our Sun dies and there is no aether to affect planets' orbits and global weather. At present scientists know too little about the nature science relating to the global climate changes and their wrong scientific concepts are harming our chance of surviving in future. They need to learn the newly expanded PM and only with correct scientific knowledge and concepts they can come up with the best scientific solutions for our welfare and future.

Scientists need to know that the atmosphere of NEU is the aether they have been looking for centuries and the common scientific origin of all universal phenomena. The roles the atmospheric NEU plays to affect global temperature are too important to be continuously ignored. For example, it is responsible for the heat energy Earth constantly produce itself to have a hot core, volcanic activities, etc. Fluctuation in the energy level of the atmospheric NEU surrounding Earth should directly affect the amount of heat energy the Earth constantly produces, its ocean temperature, and, thus, its weather and global temperature.

Scientists should also know that our Sun has at its center a DMO, which is fast rotating constantly producing a spiral neutrino wind around it. This wind produces directional antigravity to maintain the steady-state orbiting and rotational motions for all planets. Any change, gradual or sudden, in the rotational velocity of the DMO will significantly affect the motions of all planets! It is a matter of time that the change will occur and when such change occurs, it will likely end the civilization on Earth! We have found that Mars might have water and biological lives before but something has happened, likely a change of its obit occurred leading to drastic change its global temperature. It is also a matter of time that the solar system will capture another planet or even a large asteroid with or without any direct collision. The solar system is in a very dedicated equilibrium that must have taken a long time for the entire planets to gradually change their orbits to reach a final equilibrium with all planets orbiting on the same plane. When a large size object enters solar system, it will suddenly affect the orbits of all planets to gradually reach a new equilibrium system, which will cause significant weather change of the Earth. A large size asteroid had hit Earth and caused the extinction of dinosaurs, which also should have significantly altered Earth's orbit and its weather.

Earth's weather is also affected by any change in Sun's surface temperature and activities. Our Sun is a mature regular star, which periodically increases its surface activity to release its constantly building up inner-gas-pressure, which will eventually cause the outer-layer of Sun to gradual or violently expand someday. Since both big bang theory and current mathematical model are wrong, we really do not know when this will happen. Any of the above discussed possible natural causes will potentially alter the global temperature so much that all biological lives on Earth will be wiped out.

The atmosphere of NEU is also closely related to weather problems. For example, both cloud and its electric charges play important roles in producing such violent weather as hurricanes and tornadoes. The electric force is produced by NEU and they play an important role behind the destructive power of these violent weathers.

We are very unique in the entire universe and our existence relies entirely on that the solar system remains unchanged and that our Sun provides us constant amount of energy all the time. The nature is so unfriendly that we cannot survive in 99.99…% of the environment in the universe! Unavoidably many natural factors will change global weather gradually or suddenly threatening our very existence sooner or later. Upon aging the gravitational strength of Sun should get stronger while its spiral neutrino wind or directional antigravity should get weaker even when the rotational velocity of the DMO of our Sun remains constant. As a result, all planets should slowly get closer to Sun while their orbiting velocities should gradually slow down. The mainstream of thought has led us to naively believe that there is still billions of years left for the present Earth-Sun relationship or our future. Therefore, the above manmade factor has become the only urgent scientific issue.

A non-specific solution to global temperature change problem

We must rely on both Earth and Sun to survive since the nearest stars is too far (about four light years) away and they are likely unable to support a planet or even a spaceship to orbit around since they may not have a fast rotating DMOs at their centers! To prolong the existence of human race we need a viable scientific method to solve nature-caused changes on Earth temperature. The only possible way to do so is to modify Earth's orbit round Sun, for example, pushing Earth away from the Sun would solve global warming problem.

Scientists may laugh at the author's proposal since they do not believe that it is possible to modify Earth's orbit. They think that producing any force to push Earth on Earth is canceled by its own reaction force. However, they are likely wrong since gravity is a pushing force of the atmospheric NEU not an attraction force of the Earth. It means that when the solid Earth (not including its air in this case) applies a pushing force on air, air has a reaction force to push Earth in the opposite direction. Therefore, in theory we can modify Earth's orbit by pushing Earth against its air, for example, having powerful fans installed along the Tropic of Capricorn. When we turn on the fans at daytime they slowly push Earth away from the Sun and when we turn them on during the night, they push Earth towards the Sun. The force of the fans may be too small to produce significant change in Earth's orbit in hours and even days but in years it may. An international effort to establish such a capability makes great sense. We may not want to change Earth's orbit just to solve the present global warming problem. However, having such a capability will save mankind someday from all possible global temperature changes caused by many possible nature factors cited above.

Particle physics

1. According to modern physics

Particle physics had been very simple. Electron and proton were the only two fundamental charge particles serving as the building blocks of all MAM. However, physicists have found from both the debris of cosmic ray and high-energy collision experiments many other uncommon charge particles such as positron, muon, pion, etc. In fact, they have already found about a thousand of them and are building more powerful particle accelerators to find more.

Modern physics however has developed a theory, the so-called standard model, to interpret them. It is a force-field based mathematical theory having all the postulations of both theories of relativity and quantum theory and many of its own. According to theoretical physicists, the Standard Model falls short of being a complete theory of interactions, primarily because of its lack of the inclusion of gravity, the fourth known fundamental interaction.

Modern particle physics research is focused on subatomic particles, which have less structure than atoms. These include atomic constituents such as electrons, protons, and neutrons (protons and neutrons are composite particles, made up of quarks), particles produced by radiative and scattering processes, such as photons, neutrinos, and muons, as well as a wide range of exotic particles.

Strictly speaking, the term particle is a misnomer because the dynamics of particle physics are governed by quantum mechanics. As such, they exhibit wave-particle duality, displaying particle-like behavior under certain experimental conditions and wave-like behavior in others (quantum field theory). Following the convention of particle physicists, "elementary particles" refer to objects such as electrons and photons, with the understanding that these "particles" display wave-like properties as well.

All the particles and their interactions observed to date can almost be described entirely by a quantum field theory. The Standard Model has 40 species of elementary particles (24 fermions, 12 vector bosons, and 4 scalar bosons), which can combine to form composite particles, accounting for the hundreds of other species of particles discovered since the 1960s. The Standard Model has been found to agree with almost all the experimental tests conducted to date. However, most particle physicists believe that it is an incomplete description of nature, and that a more fundamental theory awaits discovery. In recent years, measurements of neutrino mass have provided the first experimental deviations from the Standard Model.

The particles of the standard model are organized into three classes according to their spin: fermions (spin-½ particles of matter), gauge bosons (spin-1 force-mediating particles), and the (spin-0) Higgs boson.

(1) Particles of matter

All fermions in the Standard Model are spin-½, and follow the Pauli Exclusion Principle in accordance with the spin-statistics theorem. Apart from their antiparticle partners, a total of twelve different fermions are known and accounted for. They are classified according to how they interact (or equivalently, what charges they carry): six of them are classified as quarks (up, down, charm, strange, top, bottom), and the other six as leptons (electron, muon, tau, and their corresponding neutrinos).

Pairs from each classification are grouped together to form a generation, with corresponding particles exhibiting similar physical behavior. The defining property of the quarks is that they carry color charge, and hence, interact via the strong force. The infrared confining behavior of the strong force results in the quarks being perpetually bound to one another forming color-neutral composite particles (hadrons) of either two quarks (mesons) or three quarks (baryons). The familiar proton and the neutron are examples of the two lightest baryons. Quarks also carry electric charges and weak isospin. Hence they interact with other fermions electromagnetically and via the weak nuclear interactions.

The remaining six fermions that do not carry color charge are defined to be the leptons. The three neutrinos do not carry electric charge either, so their motion is directly influenced only by means of the weak nuclear force. For this reason neutrinos are notoriously difficult to detect in laboratories. However, the electron, muon and the tau lepton carry an electric charge so they interact electromagnetically, too.

(2) Force mediating particles

Forces in physics are the ways that particles interact and influence each other. At a macro level, the electromagnetic force allows particles to interact with one another via electric and magnetic fields, and the force of gravitation allows two particles with mass to attract one another in accordance with Newton's Law of Gravitation. The standard model explains such forces as resulting from matter particles exchanging other particles, known as force mediating particles. When a force mediating particle is exchanged, at a macro level the effect is equivalent to a force influencing both of them, and the particle is therefore said to have mediated (i.e., been the agent of) that force. Force mediating particles are believed to be the reason why the forces and interactions between particles observed in the laboratory and in the universe exist.

The known force mediating particles described by the Standard Model also all have spin (as do matter particles), but in their case, the value of the spin is 1, meaning that all force mediating particles are bosons. As a result, they do not follow the Pauli Exclusion Principle. The different types of force mediating particles are described below.

a. Photons mediate the electromagnetic force between electrically charged particles. The photon is massless and is well-described by the theory of quantum electrodynamics.

b. The W^+, W^-, and Z gauge bosons mediate the weak interactions between particles of different flavors (all quarks and leptons). They are massive, with the Z being more massive than the $W^\pm$. The weak interactions involving the $W^\pm$ act on exclusively left-handed particles and right-handed antiparticles. Furthermore, the

$W^{\pm}$ carry an electric charge of $+1$ and -1 and couple to the electromagnetic interactions. The electrically neutral Z boson interacts with both left-handed particles and antiparticles. These three gauge bosons along with the photons are grouped together which collectively mediate the electroweak interactions.

c. The eight gluons mediate the strong interactions between color charged particles (the quarks). Gluons are massless. The eightfold multiplicity of gluons is labeled by a combination of color and an anticolor charge (e.g., red–antigreen). Because the gluon has an effective color charge, they can interact among themselves. The gluons and their interactions are described by the theory of quantum chromodynamics.

(3) The Higgs boson particle

The Higgs particle is a hypothetical massive scalar elementary particle predicted by the Standard Model, and the only fundamental particle predicted by that model which has not been directly observed as yet. This is because it requires an exceptionally large amount of energy and beam luminosity to create and observe at high energy colliders. It has no intrinsic spin, and thus, (like the force mediating particles, which also have integral spin) is also classified as a boson.

The Higgs boson plays a unique role in the Standard Model, and a key role in explaining the origins of the mass of other elementary particles, in particular the difference between the massless photon and the very heavy W and Z bosons. Elementary particle masses, and the differences between electromagnetism (caused by the photon) and the weak force (caused by the W and Z bosons), are critical to many aspects of the structure of microscopic (and hence macroscopic) matter. In electroweak theory it generates the masses of the massive leptons (electron, muon and tau); and also of the quarks.

As of September 2008, no experiment has directly detected the existence of the Higgs boson, but there is some indirect evidence for it. It is hoped that upon the completion of the Large Hadron

Collider, experiments conducted at CERN would bring experimental evidence confirming the existence of the particle.

According to the expanded PM

Using instruments having magnetic-field such as mass spectrometers scientists have successfully quantifying the masses and the charges of electron, proton, atomic nuclei, atomic and molecular ions, etc. Using instruments having magnetic field made of MAM materials to study charge particles is considered well-established but only for those charged particles made of protons and electrons. However, using them to study uncommon charge particles has not been proven valid including the interpretations of their findings! Since all of the uncommon charge particles have been found very unstable, none of the over a thousand uncommon charge particles have been isolated, and no compound has been found made of them. The entire standard model particle physics is essentially based on the ghost images detected in magnetic field.

The high-energy collision interactions inside stars would produce a thousand and more charge and non-charge particles similar to the high-energy particle accelerators. It is likely that protons and electrons can be smashed into their own AUPs, even broken into smaller, and/or even combined to form bigger AUPs. NEU may interact with some of them to form charge particles and some may remain uncharged. However, the overall findings in the past century including those generated by the high-energy particle accelerators have confirmed the very facts that all MAM are made of protons and electrons and that all charge particles found besides protons and electrons are very unstable and do not interact with one another to form stable compounds.

The conclusion from all findings is that there are only two stable FCPs, which make all MAM. A FCP is made from an AUP having an amazingly dense atmosphere of NEU. A FCP have two fundamental forces, the gravitational force or the atmospheric pressure of the NEU of the universe and the repulsive forces of its own atmosphere of NEU pushing other FCPs away, to show its charge properties. With their charge properties FCPs interact with each other to form atoms and molecules. A FCP is not further dividable and is certainly not made of quarks, et al. It is possible that their AUP are crashable to form smaller and/or even bigger AUPs but findings showed that the crashed AUPs do not form stable FCPs with NEU. FCPs are therefore the smallest units of matters. It is a dream comes true that particle physics can be so simple that everything is made of only three fundamental particles: NEU and the AUPs of electron and proton!

Light-NEU are NEU having spin energies emitted from electrons. The atmospheric NEU of an electron has the spin energies stored covering the entire spectrum of light from X-Ray to microwave. Upon collision interacting with MAM, light-NEU constantly convert to undetectable NEU, which interact with MAM to produce gravity. The undetectable NEU also interact with AUPs to form the two fundamental FCPs, which in tern form atoms and molecules. The atmospheric NEU is responsible for producing all universal forces. The strong nuclear force is its atmospheric pressure. The weak force is simply due to the presence of the atmospheric NEU and its constant bombardments make all particles such as cosmic particles and some atomic nuclei unstable. Of course, undetectable NEU interact with FCPs to produce electric and magnetic forces too.

The colission interactions of NEU with FCPs are those among NEU themselves, which are effective interactions since FCPs have amazingly dense atmosphere of NEU. Contrary to general belief, although this type of interactions produces garvity and other universal phenomena, it does not produce heat. However, the other type of collision interactions, among FCPs and MAM, produces heat as expected. This type of collision interactions is well known including all physical, chemical, and nuclear reactions. This type of collision interactions involve releasing and/or absorbing of light-NEU, thus, producing heat or absorbing heat energy. This is because that the NEU content of FCPs, particularly of electron, varies with the environment they are in in order to reach a balance of charge forces. Realizing the fundamental structure and composition of FCPs therefore leads to the understanding of all collision interactions of material particles.

To explain universal redshift of star lights, gravitational redshift of light, and the light-bending effect by gravitational force, the author postulates the presence of an atmosphere of mini-aether having masses much smaller than those of NEU. They have similar properties to NEU, are part of the composition of the FCPs, and they participate in all collision interactions like NEU. The mini-aether produces gravitational and drag effects on both light-NEU and undetectable NEU.

Life-threatening non-charged particles

Neutron bomb is made based on the fact that non-charged particles significantly larger than NEU have strong penetration power and they are life threatening. The AUPs of both FCPs have much larger sizes and masses than NEU and they are the only two AUPs interacting with NEU to form stable charge particles. The question is whether any significant amount of these AUPs and other AUPs significantly more massive than NEU constantly present? They have such unimaginably high mass density that nothing can effectively stop them. The experiments designed to detect NEU (neutrinos) deep underground use very sensitive light detection device assuming that some high energy neutrinos can occasionally knock hydrogen or other element off molecules to produce faint light. Such a detector is non-specific and also detects all AUPs more massive than neutrinos! It is therefore not sure whether we have been detecting only the AUPs significantly larger than NEU or both since NEU may be too small to knock any atom off molecules to produce light. Even if we have been detecting 100% AUPs significantly more massive than neutrinos, their amount found is too small to threaten the very existence and even the health of biological lives.

Similar experiments may be needed to routinely monitor the environment on the surface of the Earth where people live. It is similar to using Geiger counter to monitor hazardous radioactive materials to detect the levels of environmental live-threatening non-charged particles including AUPs.

Beta decay

According to modern physics

Early studies of beta (electron) decay revealed a continuous energy spectrum up to a maximum, unlike the predictable energy of alpha particles. Another anomaly was the fact that the nuclear recoil was not in the direction opposite the momentum of the electron. The emission of another particle was a probable explanation of this behavior, but searches found no evidence of either mass or charge change. Pauli in 1930 proposed the existence of a particle, which was named neutrino later, to carry away the missing energy and momentum. Until 1953 neutrino has been experimentally detected still indirectly.

Beta particles are the electrons from the nucleus, the term "beta particle" being an historical term used in the early description of radioactivity. The high energy electrons have greater range of penetration than alpha particles, but still much less than gamma ray. The radiation hazard from beta ray is greatest if they are ingested. According to current theory, beta emission is accompanied by the emission of an electron antineutrino which shares the momentum and energy of the decay. The emission of the electron's antiparticle, the positron, is also called beta decay. Beta decay can be seen as the decay of one of the neutrons to a proton via the weak interaction. The use of a weak interaction Feyman Diagram clarifies the process. In summary, the emission of a positron or an electron is referred to as beta decay. The positron is accompanied by a neutrino. Positrons are emitted with the same kind of energy spectrum as electrons in negative beta decay because of the emission of the neutrino.

In 1933 Carl Anderson found positron in the debris of a cosmic-ray collision, which had been predicted by Dirac in his theory of relativistic quantum mechanics developed two years early. According to modern physics positron is the antiparticle of electron carrying opposite charge, magnetic moment, the same mass and spin as electron. It also teaches that positrons quickly annihilate with electrons to completely turn into large amount of pure energy obeying Einstein Equation. In the high-energy accelerator laboratory some radioactive isotopes, such as carbon-11, potassium-40 can be made, which undergo the so-called positron-emission type of beta decay. This technology has been found having important medical applications such as making radioactive tracers and taking medical images, the so-called positron emission tomography (PET).

According to the expanded PM

There are enough scientific evidences to show the existence of positron although it is too unstable and its life span is too short to be collected and stored. However, the expanded PM does not support antiparticle and the current positive and negative charge concepts. It therefore needs to offer a scientific explanation why positron exists.

According to expanded PM the electrons inside atomic nuclei have lost a large amount of their atmospheric NEU in order to get to very close to protons to form neutrons. Therefore, an electron ejected from a radioactive atomic nucleus contains much less NEU than normal electrons. The electrons so released are the most NEU-deficient electrons, which would have strong tendency to capture NEU from their surrounding environment. The released electrons have weaker repulsing force but stronger gravitational force in comparison to normal electrons and as a result it may become positive in a magnetic field. It should be very unstable by capturing environmental NEU and even to capture NEU from other electrons and protons involving emitting light and heat energies. Therefore, a positron is likely a highly NEU-deficient electron freshly ejected from a radioactive atomic nucleus. Finding positron proves that electron has an atmosphere of NEU and that it has lost significant amount of NEU to form neutron inside atomic nuclei. Positron quickly becomes electron upon capturing environmental NEU, but not to annihilate with an electron.

The electrons forming neutrons with protons in atomic nuclei have lost significant amount of its atmospheric NEU. The degradation of a neutron in the atomic nuclei therefore involves releasing highly NEU-deficient electrons, which should quickly capture environmental NEU to become normal electron. The newly released NEU with high deficiency of NEU may behave like a positron. Why some beta-decays emit electrons and some positions? It may be related to the rate of radioactivity or the availability of environmental NEU for the released NEU-deficient electrons to quickly capture them. Fast beta decay process should produce positions while slow beta decay allows released NEU-deficient electrons to capture NEU to be detected as normal electrons.

According to the expanded PM, beta decay is caused by the bombardment of the atmospheric NEU of the universe and the process involving the capturing of a lot of NEU from surrounding environment by the emitted electrons, which are highly deficient in NEU content.

Space science

The expanded PM should be the new fundamental nature science of all fields including Earth and space sciences. It is time for scientists and engineers in all fields of natural science to recognize that we live under an atmosphere of NEU, which is very interactive with everything to produce all universal phenomena such as having frictional force to slow down satellites and spaceships. Designing spaceships and planning space trips must take all the aspects of the presence of the atmosphere of NEU into account. For example, before long-distance space exploration such as sending people to Mars we need to conduct animal experiments to make sure that human can live in the atmosphere of NEU in deep space. Besides, we need to take both the antigravity force of the spiral neutrino wind of the Sun and the drag effect of the atmospheric NEU into consideration for space travel calculations.

Breakthrough in space science relies on the acceptance of the new expanded PM by the scientific and engineer communities of the space science and industry. At present only conventional fuels can be used to drive spaceship, which is limited by the amount can be carried and can not reach far into space. Spaceship needs much more concentrated and long-lasting fuels, such as radioactive materials. However, no spaceship powered by radioactive fuel has been developed and MOT teaches that it is impossible to do so.

Although NEU can penetrate through all MAM, light-NEU cannot. In fact, some materials such as mirror can effectively reflect light without significantly losing its energy. Besides, some materials such as those having double and triple chemical bonds can effectively stop or absorb certain light energies from their collisions with light-energy-carrying NEU. Since light energy is carried by individual NEU all the way, light should have a pushing force on those materials capable of effectively reflecting and/or absorbing light. The author challenges scientists and engineers to develop a spaceship with its engine powered by light-energy using radioactive materials as fuel.

There is an atmosphere of NEU to produce all universal phenomena and the wrong concepts of MOT and theories have to be corrected. It is impossible to understand Earth and space science without understanding the vitally important roles the atmospheric NEU play. For example, it plays vitally important roles in the motions of the Sun and all its planets and in affecting weather, volcanic and earthquake activities.

Chemistry, biology and medical science

The progress of the natural science in all other fields has not been seriously hindered by the wrong concepts of physics and cosmology since they have long been based on PM with experimental findings to discover physical origins. They are so-far essentially immune from the mathematically derived pseudo-physics. However, since physics is the foundation of all fields of nature science, wrong physics has so far hindered the fundamental understanding of physics and the natural science in all fields. For example, the old positive-negative charge concept has hindered the scientific understanding of atoms and molecules in chemistry and biology. Fundamental understanding of the structures and the compositions of protons and electrons is important for chemists and biologists to understand charge energy, atoms and molecules. They also need to know the important role the atmospheric NEU play such as that the very existence of MAM relies on their constant collision interactions with the atmospheric NEU of the universe to continue to gain their energy. For example, besides breathing air the atoms and molecules of our bodies, plants and everything on Earth need to constantly breathe NEU to survive.

In the presence of an atmosphere of NEU question should be raised on what are the effects of change in the energy level of the atmospheric NEU on physical, chemical and nuclear reactions and biological lives. For example, if the change affects the rate of nuclear reactions, atomic clocks are no longer valid for measuring absolute time. Scientists would be interested in finding both the beneficial and the harmful effects of changing the energy level of atmospheric NEU on biological lives. We have already found that biological lives such as human can tolerate significant changes in environmental energy level of the atmospheric NEU. For example, we have found that we can live in both lower-energy NEU (deep under ground) and higher-energy NEU (space-station) environments for a short period of time. However more studies are certainly needed to utilize the benefits and to avoid the harms of these energy changes. The author also suggests developing a detection method similar to that used by Super-Kamiokande Laboratory to detect neutrinos to routinely detect the level of absolutely uncharged particles (AUP), particularly, in space, which may be related to the development of such deseases as bone loss and cancers.

The newly-discovered expanded PM will undoubtedly bring new scientific understanding, breakthroughs and new challenges to all fields of nature science.

Scientific revolution

To begin with natural science is inseparable from religion and powerful church taught it based on religious belief. In 16th Copernicus started a secret scientific revolt against the teachings of cosmology of the powerful church. A century later his follower Galileo published his observatory findings to openly support Copernicus' cosmological view. He has further led to the use of experimental findings to discover natural laws, to prove, and to disprove scientific views, The Galilean physics of mechanics (PM) he has led to develop has been repeatedly proven correct over four centuries and has long become the foundation of all fields of natural science.

In the beginning of 20th Century Einstein again has led a "scientific revolution" to launch his theories of relativity against aether concepts. Most people do not realize that his "revolution" is against the teachings of PM that phenomena have physical origins. It is the fundamental teaching of PM supported by well established findings. Upon accepting Einstein's theories of relativity, the scientific community has mistakenly recognized and taught two controversial physics in the past century!

Einstein's revolution cannot be successful since it has neither disproved PM nor proved the correctness of the postulations of his theories of relativity. Yet, he has won both official and public support leading to the development and acceptance of modern physics and modern cosmology know as the MOT. Over the past century, a an authority system, including refereed journals and national research funding organizations, has long been established using the knowledge and the formal science philosophy of MOT to suppress opposing scientific views including those of PM. Many scientists have openly called the MOT a religion and many books have also been written to detail the facts. For the public Einstein has long become the symbol of science and with both religion-like belief in Einstein and whole heartedly trust of our government, the public is firmly on the side of MOT! Unfortunately, the real physics, our welfare and future are not.

The scientific revolution of both Copernicus and Galileo has successfully separated physics and cosmology from religion leading to the use of experimental findings to discover natural laws. To bring science back to religion, Einstein's revolution is a counter scientific revolution!

Chapter 6 – Natural science and governments

Governments controlling natural science R&D

The scientific revolution of Copernicus and Galileo has led to both clear definition what natural science and scientific method are and the separation of natural science from religion. Ever since, R&D in natural science had led to industrial revolution and great improvement of living standard. The huge benefit the knowledge of natural science brings, particularly, in national defense and economy, has led most nations to invest heavily in their scientific R&D leading to worldwide competition and cooperation in scientific R&D efforts. The governments of all nations have long replaced church to be the authorities to oversee the development and the teaching of natural science. In the past century United States of America, the most scientifically advanced country in the world, has led to the development of the so-called the mainstream of thought (MOT) including both modern physics and modern cosmology based mainly on Einstein's theories of relativity.

Calling it MOT should mean that it is still an unproven theory, which may be correct or wrong. This is very true since most of the fundamental postulations of MOT, such as the invariance of light speed, the existence of both (matter-free) force fields and spacetime, have not yet been proven correct. However, MOT has long been taught in schools and colleges and its knowledge has long been used to determine what scientific papers should be published and what R&D projects should be funded. MOT therefore has long been treated by both the governments of all nations and the public as the proven-correct physics and cosmology instead of an unproved theory!

Most governments, such as the government of USA, are obligated to engage in, to fund basic scientific R&D, and to maintain the health of scientific community. They however have made a big mistake by exclusively supporting MOT in the past century using its unproven knowledge and formal-science philosophy to reject other scientific views particularly those of PM, which have been repeatedly proven correct over the past four centuries. The support by the governments worldwide has made MOT the most powerful religion-like organization in the world draining the R&D resources and putting the welfare and the future of mankind at great risk. The damage however has largely been confined within physics and cosmology since all other fields of the natural science are still based on PM. Although their support of MOT in the past century has done incalculable harms, the findings and the knowledge obtained from the wrongly directed R&D efforts still have many unintended and useful applications. For example, the knowledge and the technologies obtained from the costly high-energy acceleration R&D have found important medical applications. Besides, the development of technical applications such as inventions has been based on PM and has been flourished over the past century.

However, scientific education has suffered the most damage from the huge mistake all the governments have made. Several generations have already been misled by the wrong scientific education into believing in MOT like a religion. It is particularly easy to damage young minds from trusted wrong educations. Today, the majority of physicists, cosmologists, and even astronomers have accepted MOT and its mathematical equations to represent such important universal phenomena as light, gravity, electric charge energy, the universe, etc. and have given up logical and common-sense understanding. As expected, the public has blind trust and support of their governments in educating them and their children. They have been educated to believe that mastering difficult mathematics is necessary for understanding both physics and cosmology. Without strong background of mathematics, they have to blindly trust what the authority teaches them.

It should not be surprised to discover that PM, which has been repeatedly proven correct in the past several centuries, is the only correct physics, not MOT. Upon the discovery of the expanded PM, MOT has been disproved. In the past decade the author has been trying to convince our government and the public that the modern physics and the modern cosmology developed in the past century are wrong and that the newly-discovered expanded PM should replace MOT. Unfortunately, all his efforts have failed. He hopes that this book can convince some to continue his fight to save the only real physics. After all, it is everybody's fight for their own welfare and future and he wishes the best of mankind.

Author's fights for real physics and cosmology

As expected, the expanded PM the author has discovered has been repeatedly rejected by scientific journals for publication and by scientific authorities for research and educational funding. The physics professors and the officers of many leading universities have ignored his letters regarding this vitally important scientific discovery and the fatal scientific mistake of MOT. Finally, the author has no choice but to publish his breakthrough scientific discovery himself to directly appeal to the public.

The author's efforts to publish the expanded PM in peer-reviewed journals and to apply for research and education funding in the past decade have all failed. It was partly his own fault since in the past his scientific views were incomplete and he was unable to answer some questions raised. It was however clear that he has been in the correct track trying to explain all universal phenomena logically and consistently in terms of their only possible physical origins. It however has been painfully clear that the scientific authorities are in the wrong side of the real physics and cosmology and that they are not open minded. This must be the worst situation for the public that the most trusted scientific organizations and their authorities established to support and to advance real natural science are working against it!

1. Physical science journals

The author's efforts to publish his breakthrough discoveries have encountered unimaginable difficulties. All his submissions have been rejected by the editors of the journals of physical science, cosmology, and astronomy without giving any valid scientific reason, thus, without even having a chance to be evaluated by reviewers. There were some reasons vaguely stated by the editors either in their rejection letters or in their response to the author's persistent inquiries. For example, Physics Review Letters cited the lack of mathematical theory, the qualitative and the "speculative" nature of the paper and Canadian Journals of Physics simply cited difference in philosophy.

Physics Essay is the only journal having author's paper evaluated by reviewers. However, difficulties had developed when reviewers rejected the paper without giving valid scientific reasons. A total of five reviewers had been called upon and only one has recommended publishing it. The editor had finally rejected the paper a year after the submission. Even the only reviewer favored this paper has serious doubt about the teaching of PM – physical origins for all phenomena. His comments are summarized below:

"First, I think it is highly praiseworthy to seek to restore a physically explanatory dimension to physical science. This paper is also rich in interesting and imaginative solutions to problems." … "I support any effort to produce real physical explanations as preserving independence of thought and philosophical values against the main contemporary current. When such an effort goes forward under the materialist ontology, I see contortion and strain, the need for new ad-hoc postulates, and so on. For whatever it is worth, my own belief is that a quest for fundamental physical explanations without a shift to a post-materialist conception of physical reality is destined for failure." … "The author has satisfactorily corrected the problems of a philosophical nature that I pointed out. Assuming there are not serious flaws of a scientific nature which are outside my expertise, I think it is valuable to keep alive the explanatory questions this paper pursues, even though I disagree that a revival of classical materialism is the way to go. Therefore, because of its originality and inquiry-stimulating properties, I recommend this paper be published."

Now a day, it is hard to find an open-minded MOT supporter. The comments of this reviewer reveal how far apart the state of minds of today's physicists from the Galilean physics. They religiously believe in mathematical origins and are unwilling even to trust and to evaluate findings supporting physical origins anymore. Scientific journals have legal right to reject publishing any submitted paper with or without valid scientific reason. They have long been abusing their freedom to support the MOT like a religion!

2. National Science Foundation

As the major funding source for basic scientific research and education, National Science Foundation (NSF) plays a leading role in the scientific community. Unlike scientific journals and magazines, NSF should be officially responsible for maintaining a healthy scientific community, finding and advancing the only real natural science. NSF should obey law not to fund religion including religion-like activities. To perform its duties NSF must be able to differentiate natural science from religion and religion-like pseudo-science. It is really bad for NSF to support wrong physics intentionally or not, particularly religion-like pseudo-physics, and to deny real physics like what NSF has been doing in the past century!

In the past decade the major effort of the author has been seeking the recognition and the support of NSF of the expanded PM. The failures of his effort have been part of his fault due to the lack of completion of his proposals to NSF. However, NSF should be responsible for both the blind support of MOT and the lack of an open mind by its science officers. For example, the author has pleaded in his letters NSF to evaluate his proposals using the knowledge of PM not MOT but has been ignored by NSF officers. As the leading nation of the world, NSF has played the major role in creating an era of religion-like pseudo-physics and pseudo-cosmology in the past century!

NSF uses a peer-review method to evaluate proposals for funding, which usually does not give the authors of the proposals a chance to rebut reviewers' comments. Although it may be the best method available to handle massive amount of proposals, it has not been working due to that both the officers of NSF and the "scientific community", which the NSF chooses reviewers from, blindly support MOT against real physics! To ensure this or any other method working, the officers of NSF have to be open minded and have the fundamental scientific training to be able to judge generally the validity of scientific views based on well-established scientific knowledge and method. For example, they should know that MOT have many unproved postulations and that PM has being repeatedly proven correct for centuries and the usage of the unproven knowledge of the MOT to disprove the teachings of PM, such as that all phenomena have physical origins, should not be valid. Demanding to have a mathematical theory to back up the discovery of a physical origin of a universal phenomenon should be an unscientific requirement. At least, before PM is disproved, NSF should support both MOT and PM.

Author's first proposal to NSF

The author has submitted his first research proposal entitled "A New Universal Theory – further development and evaluation" in early 1999. It was a very early version of the expanded PM that showed scientific evidences proving that the atmosphere of neutrinos is the aether to produce both gravity and light. The proposal was rejected based on reviewers' comments given below.

Reviewer 1:

The proposal seeks further support to study a new approach to gravitation. Working from the perspective that special relativity and general relativity are not correct, and that field theory approaches to electromagnetism and gravity are unacceptable, the author attempts to explain a wide range of observations. The reviewer recognizes that one must constantly critique existing theories and propose new explanations; however, he feels that the current work falls short of existing theories by lacking internal consistency and predictive power.

The core of the proposed New Universal Theory is that gravitational interactions arise from the space subatomic particles, mainly neutrinos. Rather than accept the current field interpretation of gravity, the author looks for a physical origin for gravity. A wind of space subatomic particles, of which neutrinos play a large part, is thought to blow through the universe at very high speeds. This theory suggests that particles lose some kinetic energy as they pass through a massive body, and hence, the gravity wind flow out of the earth is weaker than the wind descending onto the earth from the heavens; this differential leads to an overall pushing of other matters towards the massive body.

The New Universal Theory is applied to two other physical situations. The first is the case of light and its bending around massive objects. Rejecting a field interpretation of electromagnetic waves, the proposal suggests that a suitable candidate for the ether is a wind of neutrinos, and that light waves are essentially mechanical in nature. Further, from the arguments given above demonstrating that a stronger gravity wind of neutrinos blows towards the massive object than away, one concludes that light will deflect towards a massive object because of the motion of the gravity wind through which the light travels. The New Universal Theory is also applied to cosmology; here, the neutrinos are assumed to have a natural dispersion which explains the expansion of the universe.

While recognizing that new formulations of gravity theories may be useful in defining a quantum theory of gravity or increasing our understanding of the world, the reviewer does not feel that the current proposal has great promise for adding to the body of knowledge. Granting for a moment that the New Universal Theory provides an internally consistent world view, it is not clear that the perspective of the theory has the potential to describe the wide range of observations currently explained by general relativity. In addition, the proposed theory do not (yet) give any sort of prediction, which can be tested to see if it fits our observations any better than existing theories.

Regarding the proposed New Light Theory, the reviewer wonders if the author is familiar with the work of many prominent scientists at the end of the 19th century on the properties of ether. A solid check of internal consistency of the New Universal Theory would be to show that a collection of neutrinos satisfies the properties which the ether must posses to transmit electromagnetic waves.

The reviewer sees that the proposal has the possibility of being an important step in our understanding of the world. However, the current proposal is lacking in two ways. First, the reliance on neutrino as a gravity wind by the New Universal Theory seems to limit the power of the theory to explain all of the observations explained by existing field theories. Second, the proposed theory has yet to make any predictions which could be tested for accuracy. The reviewer feels that both of these issues should be addressed by the author of the proposal.

Reviewer 2:

The following report is based on the March 26, 1999 revised version of 'New Universal Theory', which is part of NSF proposal PHY-9907068. In support of his proposal, Dr. Tsau presents a 'New Universal Theory', which is based on the assumption that 'essentially undetectable' subatomic particles occupy all of space. These 'space subatomic particles' (SSPs) are used by Tsau to develop new theories of gravity, light, and the expansion of the universe. Tsau claims that the existence of SSPs disproves Einstein's theory of special relativity and extends the general theory of relativity, by providing an underling mechanism that is responsible for the transmission of light and gravitational interactions. I am not convinced, however, by Tsau's arguments, and I do not recommend that the proposal be funded.

1. Although Tsau claims that a 'gravity wind' of SSPs is the mechanical explanation of the gravitational fields associated with massive objects, this "New Gravity Theory' cannot explain, for example, how the earth could remain in orbit around the sun for billions of years without slowing down due to collisions with the SSPs. (More SSPs will hit the front of the earth than the back as the earth moves in its orbit, creating a force that opposes the motion.) In addition, Tsau states that 'for very dense matter, the strength of gravity is affected by its size only and no longer affected by the amount of its mass.' This is direct contradiction to experimental observations which show that the gravitational held of a massive object is proportional to its mass and not its size.

2. When discussing the Doppler Effect of light, Tsau states that 'if there is no transmission medium, conventional Doppler Effect no longer exists. A Doppler-like effect can only be observed by changing the speed of emitted particles'. This statement is incorrect. In special relativity, even though one assumes the absence of a transmission and an absolute, observer-independent speed of light, one is still able to predict the existence of a Doppler Effect when there is relative motion between the receiver and source of light. More over, this predicted Doppler Effect has been verified, for example, by the red-shifted light we observe from receding galaxies. Tsau also fails to realize that the Doppler Effect for sound is fundamentally different from that for light ---i.e., the

Doppler Effect for light depends ONLY on the relative motion of source and receiver, while that for sound depends on whether the source or receiver is moving with respect to the transmission medium.

3. According to Tsau's 'New Light Theory', light travels 'at a constant speed … (only) relative to its transmission medium, SSP. If light is emitted by an (outer) galaxy which is approaching or receding from us, the speed of light would be correspondingly larger or smaller. In making such statements, Tsau is incorrectly applying Galilean addition of velocities to relativistic objects (such as light). The fact that the Galilean addition of velocities formula fails for relativistic objects (and should be replaced by the Lorentzian addition formula) is verified experimentally everyday in high-energy accelerators. Tsau seems to have neglected this observational fact when formulating his theory.

4. Tsau claims that current theories of gravity do not obey the law of conservation of energy. This statement is incorrect. Newtonian gravity and Einstein's theory of general relativity DO obey the law of conservation of energy (albeit only globally in general relativity). Granted, Newton's theory of universal gravitation and Einstein's general relativity do not provide the type of mechanistic description of gravitation that Tsau wants, but I do not agree with his statement that general relativity is only a 'pure mathematical model without physical meaning'. A theory (like general relativity) which agrees with experiment and explains HOW objects move in response to gravitational interactions (but does not necessarily explain WHY) is better than a theory which purports to answer WHY but does not agree with experiment.

For these and other reasons not mentioned here, I do not recommend that the proposal be funded.

Author's 2nd proposal to NSF

Usually NSF does not accept re-submissions of rejected proposals or similar ones. It was because that the author wrote to the Director of NSF Dr. Colwell pleading her to save our science. Dr. Dehmer, Director of the Division of Physics, responded suggesting that I submit a proposal. My second proposal "New Aether

Science" was submitted in early 2003 and was rejected in June. The comments of reviewers are given below.

Reviewer 1: It is so differently at odds with the current standard view of physics. I think it would be inappropriate for the NSF to support this work at any level until the PI has demonstrated its credibility by generating a series of published accounts of the work in the main stream refereed scientific literature.

Reviewer 2: The proposer has no track record of published research in gravitation physics in refereed publications.

Reviewer 3: There is no published record of the investigator's previous work in the field. The ideas presented in the proposal are purely qualitative and no quantitative underlying theory apparently exists at present. ...at present string theory offers an entire gamut of new alternatives...

Author's rebuttal

The Author apologies that the expanded PM given in both of his proposals to NSF have not been as complete as he had hoped. The first proposal did not address some scientific questions raised by the reviewers and the author did not have a chance to answer them. However, the answers to all the scientific questions raised by reviewers can be found in this book. The majority of reviewers' comments cited the conceptual, the philosophical and the methodical differences between PM and MOT simply proving their incompatibility with each other.

It is very clear from the comments of all reviewers of both proposals that the real reason for them to reject both proposals is because of the incompatibility between the expanded PM and MOT. They cited that both the findings and interpretations given in the proposal did not have a (mathematical) theory to substantiate them. They rejected the proposals' "qualitative" interpretations. In short, they used the knowledge and the formal-science philosophy of MOT to reject the proposals. They also relied on that the author could not publish his discovery in peer-reviewed physics journals to reject the proposals.

If MOT were correct, it should be both consistent and coherent among all its own theories. It should also be consistent with PM since it has been repeatedly proven correct in the past three centuries. However, MOT is neither consistent within itself nor with PM. For example, it is well known that theories of relativity are incompatible with quantum theory and their postulations contradict with the teachings of PM. If curved spacetime produces gravity, to be consistent it should be applicable to interpret all forces such as electric and magnetic forces, the atmospheric pressure of air, etc. It cannot. If universal phenomena such as light and gravity had mathematical origins, all phenomena such as sound, fire, a disease, etc. should have mathematical origins too, but they do not. The mathematical interpretations of the universal phenomena of the MOT are neither consistent nor coherent among themselves. For example, all its universal forces are unrelated and are produced by different force particles and most of them have not been found present in space like NEU! MOT is still far from having a complete, consistent, and coherent interpretation of the universe and its phenomena. It has been the dream of Einstein and many others to develop a theory of everything but ended only with frustrations. Many still put their hope on theories such as the string theory under development.

Unlike other fields of natural science having essentially unified views, there have long been at least three versions of physics coexisted supported by many physicists but are contradictory with one another, such as PM, MOT, plenum aether theories, etc. The competition therefore has been on for at least over a century already for any of them to come up with an overall interpretation of the universe consistently and coherently to prove that it is the only correct or real physics. Over the past century, although the world has betted on MOT with the support of the entire R&D resource worldwide, it has not won!

Never before, a logical interpretation, which is both consistent and coherent and is supported by the well-established findings cumulated over the past four centuries, of the universe and all universal phenomena with their physical origins have been found. The discovery has proven that PM is the only correct physics, which should not be surprising since PM has been the only physics that has been repeatedly proven correct over several centuries.

Never again, an alternative interpretation can be found based on any other theories incompatible with PM! Now, all theories incompatible with PM including MOT have been proven wrong.

Finally, the competition among all the versions of physics is now over. All the criteria both the NSF and its reviewers used to reject the two proposals have been proven invalid!

Letters from NSF Officers

The following letters from NSF officers reveal their biased views supporting MOT.

From Dr. Dehmer, Director of Physics, dated on 2/14/03

Dear Dr. Tsau: Thank you for your 9 January 2003 letter to Dr. Rita Colwell expressing your concerns about the progress of modern physics and transmitting your book, "Mysteries of Nature." I am impressed with your dedication to science and your independent work. Your revival of the aether concept in a modern context is, indeed, quite a departure from current ideas. Many individuals with unorthodox ideas have had difficulty gaining acceptance in the MOT. However, the ideas are tested by criticism and experimental tests; and the best concepts, models, and theories generally emerge over time.

You obviously feel that you are in such a position now, having non-standard ideas and experiencing difficulty gaining acceptance. You appeal to the NSF because we sponsor basic research and have a responsibility for the health of science. However, our role is not to pass judgment on new ideas. Our role is to sponsor the most meritorious research efforts. The merit of proposals is determined by peer review, which is the best tool we have to assess the quality of proposals.

We will be happy to consider any research proposal that you may wish to submit. However the burden of demonstrating the correctness and promise of your ideas lies with you. This is normally done through peer-reviewed publication.

From Dr. Berger, Program Director, Gravitational Physics, dated on 3/14/2003

Dear Dr. Tsau: While some aspects at the frontiers of physics are controversial, the core of physics is based on agreement between theory and experiment. Even "revolutions" in physics such as relativity and quantum mechanics did not destroy the previous core but rather demonstrated that it had a limited region of validity. The "new" physics was adopted into the core when it became clear that it, too, was supported by experiments in the regions where the "old" core did not apply. A new revolution must also be consistent with previous experiment and should itself be testable by existing or new experiments. You should note that, in addition to textbook experiments, the current core of physics is supported by the successful functioning of modern technology. An interesting example of this is the Global Positioning System that has effects of special and general relativity included in its hardware and software. Thus, every time a position is correctly determined, special and general relativity are validated.

Comments from the physicist of the OIG of NSF

NSF has its own Office of Inspector General for appealing, which sought the opinion of another scientist of its own. After the rejection of his second proposal by NSF, the author has written to the Department of Justice (DoJ) leading to the investigation of the Office of Inspector General (OIG) of NSF. The following is the author's summary of the internal memorandum of NSF OIG obtained via a FOIT request:

The reviewers noted the complaint has no theory to substantiate his hypothesis and no testable predictions. The reviewers also noted the complainant has no publications in any peer-reviewed journals upon which they could assess his hypothesis. I conclude the complainant's NSF proposal was handled fairly.

Author's opinions

Dr. Dehmer agrees that NSF has a responsibility for the health of science. He however called the expanded PM given in the book "Mystery of Nature" unorthodox ideas apparently due to its incompatibility with MOT. However, he has ignored the fact that the physics in this book is consistent with PM and, even according to Dr. Berger, PM is still been recognized to be the core of physics. The major problem is that MOT is inconsistent with PM. To support both the knowledge and the philosophy of MOT, physicists have long denied the teachings of PM, when it is applied to interpret universal phenomena and the universe. The contradictions between MOT and PM have not surfaced in the past only because that PM has not been applied to explain both the universe and universal phenomena. The discovery of the expanded PM unavoidably has opened the contradictory issues between PM and MOT.

Since PM has been repeatedly proven correct in the past several centuries, it is a well-established correct physics. NSF should consider the possibility that MOT is wrong. At least both physics now should be treated equally but independently until a scientific choice on the correct physics can be made between the two or from a third physics. It should be unscientific for NSF to use the knowledge and the formal-science philosophy of MOT instead of the teachings of PM to judge the author's proposal, as requested by the author. Dr. Dehmer or NSF has not given author's proposals a chance to be scientifically judged.

Dr. Berger's letter explained in more detail about the official stand of NSF on the scientific issues of physics. He recognizes that modern physics or MOT as a newly added core of physics while the classical physics including PM remained to be the original core of physics. The reason given is that the new core has demonstrated to be supported by experiments in the ranges or environments where the original core fails to agree with experiments. By doing so, there are two incompatible core physics, thus, choosing one would be against the other. NSF apparently has chosen the MOT against PM by allowing reviewers to use the knowledge and the formal-science philosophy of the MOT as the scientific criteria to judge and to reject the PM-based proposals of the author ignoring his request to use PM knowledge to judge the proposal. It is not surprising that the scientist of OIG of NSF also support MOT. Although PM is still recognized as the core of physics, its teachings have long been violated upon the acceptance of MOT. If

MOT were correct, the physical origins of the universal phenomena or the expanded PM should never be discovered.

There can be only one correct physics that can offer an overall interpretation of the universe and all phenomena consistently and coherently supported by all findings. Now, based on this criterion the author has proven that PM is the only correct physics (and cosmology). It means that all physics incompatible with PM has been proven wrong.

Abbreviations

AUP – Absolutely uncharged particles such as NEU

CMB – Cosmic microwave interference background

DMO – Dense-matter objects: Having equal number of protons and electrons formed outside the universes

DON – Domain of the nature: The entire nature that host all universes

FCP – Fundamental charged particles: proton and electron

GPS – Global positioning system

IS – Intergalactic space

K – Degree in absolute temperature scale

MAM – Matters having atomic and molecular structures

MOT – The mainstream of thought. It includes Einstein's theories of relativity, big bang theory, quantum theory, modern physics, and modern cosmology.

NEU – The tiny particles constantly produced by the nuclear reactions of stars. They may be neutrinos if the current theory is largely wrong. They are also the particles proposed by Le Sage to produce gravity and proposed by Newton to produce light.

NSF – National Science Foundation

PM – Galilean physics or physics of mechanics

Made in the USA
Monee, IL
08 July 2026